MYCORRHIZA

AN ACCOUNT OF NON-PATHOGENIC INFECTION BY FUNGI IN VASCULAR PLANTS AND BRYOPHYTES

By

M. C. RAYNER
D.Sc. (London)

NEW PHYTOLOGIST REPRINT, No. 15

British Library Cataloguing-in-Publication Data
A catalogue record for this book is available from the British Library

Botany

The term 'botany' comes from the Ancient Greek word *botanē*, meaning 'pasture', 'grass', or 'fodder', in turn derived from *boskein*, meaning 'to feed or graze'. It chiefly involves the study of plant life, as a branch of biology.

Traditionally, botany has also included the study of fungi and algae by mycologists and phycologists respectively, with the study of these three groups of organisms remaining within the sphere of interest of the International Botanical Congress. Nowadays, botanists study approximately 400,000 species of living organisms of which some 260,000 species are vascular plants and about 248,000 are flowering plants.

Botany originated in prehistory as herbalism with the efforts of early humans to identify – and later cultivate – edible, medicinal and poisonous plants, making it one of the oldest branches of science. Examples of early botanical works have been found in ancient texts from India dating back to before 1100 BCE, in archaic Avestan writings (an Iranian language known only from its use in Zoroastrian scriptures), and in works from China before it was unified in 221 BCE.

Modern botany traces its roots back to Ancient Greece, specifically to Theophrastus (c. 371–287 BCE), a student of Aristotle who invented and described many of its principles. Today, he is widely regarded in the scientific community as the 'Father of Botany'. Theophrastus's major works, *Enquiry into Plants* and *On the Causes of Plants* (both looking at plant structure, variety, reproduction and growth), constitute the most important contributions to botanical science until the Middle Ages, almost seventeen centuries later. Another work from Ancient Greece that made an early impact on botany is *De Materia Medica*; a five-volume encyclopaedia about herbal medicine written in the middle of the first century by Greek physician and pharmacologist Pedanius Dioscorides. *De Materia Medica* was widely read for more than 1,500 years subsequently.

Medieval physic gardens, often attached to monasteries, contained plants of great medicinal importance. They were forerunners of the first botanical gardens attached to universities, founded from the 1540s onwards. In the mid-sixteenth century, botanical gardens were founded in a number of Italian universities – and the Padua botanical garden in 1545 is the first of such, still in its original location. These gardens continued the practical value of earlier 'physic gardens' of the monasteries, and further supported the growth of botany as an academic subject.

Botanic gardens encouraged the work of academics such as the German physician Leonhart Fuchs (1501–1566). Fuchs was one of 'the three German fathers of botany', along with theologian Otto Brunfels (1489–1534) and physician Hieronymous Bock (1498–1554). Fuchs and Brunfels broke away from the tradition of copying earlier works to make original observations of their own, whilst Bock created his own system of plant classification. In 1665, using an early microscope, another famed botanist, the Polymath Robert

Hooke (1635 - 1703) discovered 'cells' in plant tissue, a term he coined. During this early period, lectures were also given about the plants grown in the specially constructed botanic gardens, and their medical uses demonstrated. Botanical gardens came much later to northern Europe – largely due to the obvious differences in temperature. The first in England was the University of Oxford Botanic Garden, constructed in 1621.

Efforts to catalogue and describe the collections of these gardens were the beginnings of plant taxonomy, and led in 1753 to the binomial system of Carl Linnaeus that remains in use to this day. Linnaeus's system was a hierarchical classification of plant species providing a solid reference point for modern botanical nomenclature. This established a standardised binomial or two-part naming scheme where the first name represented the genus and the second identified the species within the genus. For the purposes of identification, Linnaeus's *Systema Sexuale* classified plants into twenty-four groups according to the number of their male sexual organs. The twenty-fourth group, *Cryptogamia*, included all plants with concealed reproductive parts, mosses, liverworts, ferns, algae and fungi.

Increasing knowledge of plant anatomy, morphology and life cycles led to the realisation that there were more natural affinities between plants than Linnaeus had indicated however. Scholars such as Adanson (1763), De Jussieu (1789), and Candolle (1819), all proposed various alternative natural systems of classification that grouped plants using a wider range of shared characters and were extensively followed. The Candollean system reflected his ideas of the progression of morphological complexity and were developed the classifications by Bentham and Hooker, influential until the mid-nineteenth century. Darwin's publication of the *Origin of Species* in 1859 and his concept of common descent, further required modifications to the Candollean system to reflect evolutionary relationships as distinct from mere morphological similarity.

In the nineteenth and twentieth centuries, new techniques were developed for the study of plants, including methods of optical microscopy and live cell imaging, electron microscopy, analysis of chromosome number, plant chemistry and the structure and function of enzymes and other proteins. In the last two decades of the twentieth century, botanists exploited the techniques of molecular genetic analysis, including genomics and proteomics and DNA sequences to classify plants more accurately. Particularly since the mid-1960s there have been advances in understanding of the physics of plant physiological processes such as transpiration (the transport of water within plant tissues), the temperature dependence of rates of water evaporation from the leaf surface and the molecular diffusion of water vapour and carbon dioxide through stomatal apertures.

Twentieth century developments in plant biochemistry have been driven by modern techniques of organic chemical analysis, such as spectroscopy, chromatopgraphy and electrophoresis. With the rise of the related molecular-scale biological approaches, the relationship between the plant genome and most aspects of the biochemistry, physiology, morphology and behaviour of plants can be subjected to detailed experimental analysis. Such developments have enabled advances in areas as diverse as pesticides, antibiotics and pharmaceuticals, as well as the practical application of genetically modified crops designed for traits such as improved yield.

The study of botany is an incredibly important science, as plants underpin almost all life on earth. They generate a large proportion of the oxygen and food that provide humans and other organisms with aerobic respiration with the chemical energy they need to exist. In addition, they are influential in the global carbon and water cycles and plant roots bind and stabilise soils, preventing soil erosion. Plants and the science of botany are crucial to the future of human society – allowing insight into our food, natural environment, medicine and products. It is a branch of human endeavour with an incredibly long and varied history, and it is hoped the current reader enjoys this book on the subject.

"Living things are found by a simple experiment to have powers undreamt of, and who knows what may be behind?"

W. BATESON.

PREFACE

A COMPREHENSIVE account of mycorrhiza and the facts relating to fungal infection of a mutualistic kind is overdue, and an attempt to provide it requires no apology. The literature of the subject is extensive and scattered, the opinions of workers often conflicting, and the original papers not always readily accessible for reference. Moreover, the treatment in general textbooks of Botany, necessarily limited in scope, is often misleading and suffers almost invariably from lack of a definite point of view in respect to the relation of mycorrhiza with the phenomena of symbiotic association in general.

The mycorrhizal habit first attracted attention as a characteristic of woodland soils. Modern experimental research has once more focussed attention on the significance of this fact, and emphasised the importance of mycorrhiza as a biologic soil factor affecting the vegetation of woodlands in common with that of other types of humus soil. The practical application of the newer knowledge to natural conditions has already attracted the attention of foresters thus opening up a new and fascinating field of research, promising alike to the botanist, the forester and the student of soil science.

My thanks are due to many colleagues and friends for assistance during the preparation of the book, and to the editors, proprietors and publishers of various scientific journals for permission to reproduce certain of the figures.

In particular, I am indebted to Dr E. J. Butler, Professor Percy Groom, Professor Augustine Henry, Dr J. Magrou of the Pasteur Institute and Mme Noël Bernard for gifts or loans of some of the earlier papers on mycorrhiza and for other help with the literature, to Dr Somerville Hastings and Mr W. Paulson for the original photograph reproduced in Fig. 51, to Mr A. C. Forbes, Director of Forestry in the Irish Free State, for that in Fig. 52, to Mr J. Ramsbottom for that in Fig. 25, and to Mr E. Clement for that in Fig. 24.

I have to thank Mr S. Garside for the original drawing reproduced in Fig. 63 and the material used for the photomicrograph in Fig. 58, and to Miss D. Freeman, B.Sc., and her pupil, Miss D. A. Clark, for noting the occurrence of mycorrhiza in *Aspidium* and sending me the preparations used for the photomicrographs reproduced in Figs. 59 and 60.

I have to acknowledge the courtesy of the editors and proprietors of the *Annales des Sciences naturelles* for permitting me to reproduce Figs. 14 and 15 from the late Dr Noël Bernard's paper, *L'Evolution dans la Symbiose* (1909), also Figs. 57 and 64 from those by Dr J. Magrou on *Symbiosis* (t. 3, 1921 and t. 7, 1925); likewise, that of the editors and proprietors of the *Revue Générale de Botanique* for permission to use Figs. 12 and 13 from M. I. Gallaud's papers on mycorrhiza (t. **17**, 1905).

Figs. 1 and 2 have been taken from *The Phytologist*, 1844. Figs. 5–10 from Johow, *Jahrbuch für wissenschaftlich Botanik* (Bd. **16**, 1885, **20**, 1889); Fig. 11 from Stahl, *Jahrbuch für wissenschaftlich Botanik* (Bd. **34**, 1900); Figs. 3, 4, *a*, *b*, *c* from Frank, "Ueber neuen Mykorrhiza-formen," *Berichte der deutschen botanischen Gesellschaft* (Bd. **5**, 1887); Fig. 16 from Janse, *Annales du Jardin botanique de Buitenzorg* (t. **14**, 1896–7); Figs. 22 and 23 from Professor Hans Burgeff's monograph, *Die Wurzelpilze der Orchideen* (Jena, 1909); Figs. 27 and 28 from Kusano, *Journal of the College of Agriculture*, Tokio, vol. **4**, 1911.

For the necessary permission to reproduce Figs. 50, 53, 54, 55 and 56 I am indebted to Dr E. Melin and the editors and proprietors respectively, of *Mykologische Untersuchungen* (Falck) (1923) and the *Svensk Botanisk Tidskrift* (Bd. **17**, 1923).

To Dr B. Peyronel and the editor of *Memorie della R. Stazione di Patalogie vegetale*, Rome, I am indebted for permission to use Figs. 46 and 47 from vol. **5** (1924) of that journal. Fig. 26 I owe to the kindness of Professor Lewis Knudson who lent me the original negative, and to the editors and proprietors of the *Botanical Gazette* who permitted its reproduction (*Bot. Gaz.* vol. **77**, 1924).

To the courtesy of the editors and proprietors of the *Annals of Botany* and that of the individual authors I am indebted for Figs. 43, 44, 45 (Rivett, vol. **38**, 1924), Figs. 48 and 49 (Weiss, vol. **18**, 1904), and Figs. 61 and 62 (McLennan, vol. **40**, 1926); and to that of Mr J. Ramsbottom, and the editor and proprietors of the *Transactions of the British Mycological Society* for permission to use Figs. 18–21 and the loan of the original blocks (vol. **8**, 1922).

Figs. 29, 30, 31, 32, 34 are original photographs and Figs. 33, 35, 36, 37, 38, 39, 58, 59 and 60 original photomicrographs. Permission to reproduce Figs. 36, 37, 38, 39 and 40 has kindly been granted by the editor of the *British Journal of Experimental Biology* (vol. **2**, 1925) and the Company of Biologists Limited.

Finally I must acknowledge my great indebtedness to my husband for assistance with the photomicrographs and continuous interest and help in the preparation of the manuscript and the correction of proofs.

M. C. RAYNER

BEDFORD COLLEGE,
UNIVERSITY OF LONDON.

April, 1927.

CONTENTS

THE EARLY PERIOD: 1840–1880

THE SECOND PERIOD: 1880–1900 CIRCA

THE MODERN PERIOD: 1900–1925

Contents

CHAP.

CHAPTER I

THE EARLY PERIOD

Introductory—Historical—The Early Period: 1840–1880—Schleiden—Reissek—The *Monotropa* controversy—De Bary's theory of symbiosis.

STRICTLY considered, knowledge of mycorrhiza dates from the year 1885, when Frank first applied the name to roots of trees showing a regular and characteristic infection by fungal mycelium.

Taking a broader view, the history of the subject falls naturally into three periods, the order of which coincides roughly with progressive changes in the points of view actuating research on the subject. Firstly, there is the period previous to 1880 during which a number of observations bearing on the subject were recorded, the full significance of the facts being either unappreciated at the time or becoming the subject of considerable controversy among botanists. Most of these early observations fall within the period 1840–1880.

A second period began in 1881 with the work of Kamienski on *Monotropa hypopitys*, and the researches, experiments and speculations of Frank and his fellow workers, and extended over the last two decades of the nineteenth century. This was essentially one of observation and speculation, largely dominated by the work of Frank and his school, and it passed without a break into what may be called the Modern Period, marked especially by the application of modern experimental methods of research to the problems presented by mycorrhiza plants.

THE EARLY PERIOD: 1840–1880.

Records of observations made during the first decade of this period and in that preceding it are of interest historically. Some of the earliest noted the presence and described the structure of curious thread-like bodies within the root cells of plants. In certain cases, subsequent observers identified the threads as fungal filaments and noted their regular occurrence in the roots of vascular plants; in others the real nature of the "hairy structures" associated with roots remained doubtful and aroused considerable controversy among contemporary botanists.

Among the earliest observations bearing on the subject are those of Meyen (1829), Nägeli (1842) and Schleiden (1842). The two first-named workers put on record the regular appearance of fungi within the cells of a number of plants but did not arrive at correct conclusions as to their significance or mode of origin within the cell.

Meyen's attention was attracted to the root tubercles of Alder which he believed to be parasites of similar habit to members of the Balanophoraceae and Orobanchaceae, although lower in development and systematic position. Nägeli recorded the presence of filamentous fungi in the roots of various species of *Iris*, and also in other genera,—"In diesen parenchymatischen Zellen lebten 3 verschieden Arten von Pilzen." He described intracellular flask-shaped bodies produced at the ends of the threads, and placed these root fungi in a new genus, *Schinzia*.

The Orchids were among the first plants to attract attention in respect to the anomalous structure of their root cells. Link (1840) figured cells containing fungus mycelium from the young seedling of *Goodyera procera*, but was quite ignorant of the real nature of the granular cell inclusions which he observed.

The observations of Schleiden (1842) on the roots of *Neottia nidus avis* are of particular interest as being the first bearing specifically on the subject of root infection by fungi. He described with astonishing accuracy the microscopic appearance of the various types of cell in the cortical tissues of the roots of this Orchid, their slimy contents and the presence of thread-like structures in many of the cells. Describing the structure of the threads more fully he noted that they were hollow and branched freely, and also that they sometimes formed a tangled skein the branches from which ended blindly within the cells. Lacking accurate information as to development, Schleiden expressed no opinion as to the real nature of these threads, beyond suggesting a possible analogy with the cells showing spiral markings described by Gottsche (1843) in the tissues of a Liverwort, *Preissia commutata*. With admirable frankness he thus put on record his sense of the remarkable nature of the root tissues and his ignorance of their significance to the plant:—"Ueber die Bedeutung dieser eigenthümlichen Bildungen weiss ich gar nichts zu sagen."

It was Reissek (1847) who first attempted to determine the origin and real nature of the thread-like structures present in the root cells of certain vascular plants; incidentally he attempted to correlate the observations of Meyen and Nägeli just mentioned with the results of his own more extended researches. Noting their regular appearance in the roots of many Monocotyledons and Dicotyledons, Reissek reached the correct conclusion that the thread-like structures were undoubtedly of fungal nature. He concluded, further, that the root fungi (Wurzelpilzen) reached their highest development in the underground roots of Orchids, and, although relatively common in the roots of other Monocotyledons and in Dicotyledons, were present only in a rudimentary condition in plants other than the Orchids:—"Hier sind nur die Keime desselben vorhanden."

It was mainly upon researches on Orchid roots that Reissek based his remarkable conclusions as to the origin of these intracellular fungus threads. He described their development from granules included among the ordinary cell constituents and in his view identical with the primordia of starch grains and chlorophyll grains. The granules were held to be fungus spores that developed into spindle and rod-like structures and finally to a skein of hollow threads as in the root cells of *Neottia*. Stages in this process were traced and figured in the root cells of *Goodyera discolor*. Presumably the spores originated from the granules *de novo*, since it is elsewhere explicitly stated by the author that the mycelium did not form spores within the cells but did so only when grown in the air.

It is somewhat surprising to find that Reissek attempted to extract and cultivate the intracellular mycelium present in roots. In view of the technique adopted, not so surprising to learn that among the fungi isolated were a species of *Fusiformium* (from *Orchis morio*) and species of "*Botrytis, Penicillium* and *Cladosporium*" from other Orchids. Spores of various species of these fungus genera are widely distributed both in air and in the soil; species of *Penicillium* and *Cladosporium* in particular are among the commonest constituents of the epiphytic mycelial flora of roots and appear regularly when attempts are made to isolate the true root endophytes by unsuitable methods. One or other member of the group has been recorded as a specific endophyte by many investigators subsequent to Reissek, but no satisfactory evidence has ever been provided that any species of the genera named is concerned in the formation of mycorrhiza.

This contribution by Reissek is notable as the first full and at all accurate account of the association between root cells and fungus mycelium now known as mycorrhiza.

Study of these older papers recalls the views current at the time respecting the origin of Fungi and Bacteria. By the early botanists they had been regarded not as living organisms but as *lusus naturae*.

Subsequent to their tardy recognition as true members of the vegetable kingdom, they were for long believed to be produced by spontaneous generation from inorganic material, or to arise from organic substance not in itself of fungoid or bacterial nature. This theory assumed that constituent particles of living cells belonging to the higher organisms could continue to live after the death of the body of which they had formed part, and could develop under favourable conditions into fungi and bacteria that were capable of producing germs and of giving rise to progeny specifically resembling the parents (Ehrenberg 1820).

The microzyme theory of Bécamp, a logical statement of this point of view, assumed the existence of very minute bodies—"granulations moléculaires" or "microzymes"—in the substance of plants and animals, occurring everywhere, enjoying an almost unlimited duration of vitality, and giving rise under suitable conditions to bacteria, sprouting fungi, and similar forms. The thesis embodying these views in detail was first published in Paris in 1864, reproduced in the *Transactions of the Medical Congress* at London in 1881, and republished in an extended form in Paris as late as 1883.

In England the attention of botanists was first attracted to the association of fungus mycelium with roots by a controversy respecting the alleged parasitic habit of the Yellow Bird's Nest, *Monotropa hypopitys*. In 1844 an English periodical—*The Phytologist*—published a number of contributions to a discussion on this matter, in the course of which attention became focussed upon the nature of the fibrous investment noticed upon the roots of *Monotropa*.

Luxford (1844) held that this plant obtained at least a part of its nourishment from "a layer of vegetable matter, consisting chiefly of the slowly decaying leaves of the beech, which are generally covered with a white byssoid fungus." Alone among the contributors to the discussion, Lees (1844) took the view that the plant was a true parasite. He figured a young plant of *Monotropa* growing from "its radical parasitical knob," and described in detail the hairy knobs upon the beech roots, "of all sizes from that of a pea, from which the little embryo *Monotropa* was sprouting, to that of a crab, nourishing a full company of several plants." Lees described also the hairy fibres by which the *Monotropa* plant attached itself to the root of Beech. These fibres he regarded as special organs functioning as suckers, and he noted that the ball or "nidus" at the base of the *Monotropa* plant is obscured with "a hirsuture that appears like a byssoid fungus." (Fig. 1.)

Newman (1844) regarded the fibres as essentially part of the root itself and Wilson (1844) recorded his view that they represented the matted extremities of grasses with which the *Monotropa* roots were in contact. Alluding to his examination of the roots of plants collected near Southport (a locality where *Monotropa* is still found) and believed to be parasitic upon the roots of *Salix argentea*, he wrote as follows:—

> ...the mass composing the root of the Southport *Monotropa* had a white covering of a matted and somewhat woolly substance, supposed to proceed from the radicles of the grasses which grew with it.... There was not even contact, much less parasitical connection, between this white coating and the roots of the *Salix* contained in the sod.

In view of these very conflicting opinions, Rylands (1844) re-investigated *Monotropa* material from various localities and eventually reached correct conclusions regarding both the habit of the plant and the true nature of the fibrous investment of the roots. "The plant," he states, "is not parasitic; it has no organic connection with the 'nidi' of roots among which its own are developed." From a careful examination of the fibres he concluded also that the "byssoid substance is really fungoid, and performs no essential function in the economy of the *Monotropa*."

In order to test the correctness of his observations on *Monotropa*, he examined the roots of a number of other plants and reported as follows:—"the *really* fungoid matter found on the roots of groundsel, Epilobium, Plantago, etc., had so much resemblance to the substance in question, that it would be difficult by words to render the difference appreciable."

Rylands published careful drawings of the root of *Monotropa* with its fungus sheath and figured details of the hyphae. One of these drawings (Fig. 2) is of rather special interest in relation to modern work and will be mentioned again. With the help of Berkeley—the founder of British mycology—he referred three out of the four fungi distinguished upon the roots to known genera: the other was placed provisionally in a new genus, *Epiphagos*.

Ryland's observations seem to have temporarily exhausted interest in the subject in England but the controversy respecting the parasitic habit of *Monotropa* was renewed subsequently by various continental botanists. Thus, in a work on comparative anatomy by Chatin (1856) *Monotropa* was described as a parasite in the younger stages of growth; later it became detached from its host and grew saprophytically. The accurate observations of Solms Laubach (1868) demonstrated that no organic continuity could be detected between the roots of *Monotropa* and those of the tree beneath which it grew, however close the contact might be, and, therefore, that the former plant was not a parasite. Drude (1873) again described and figured a parasitic connection between the roots of *Monotropa* and those of *Abies excelsa*, an error of observation accounted for later by Kamienski. In the same paper Drude mentioned the mycelium associated with the roots of *Monotropa*, and compared it with the appearances described long previously by Schleiden in *Neottia*.

With the exception of these controversial records concerning *Monotropa*, the possible value of observations bearing upon the association of fungi with the roots of vascular plants seems to have escaped the attention of botanists for a number of years. Isolated observations by various continental botanists continued to place on record the presence of fungal hyphae in and upon roots, although the time was not yet ripe for an appreciation of the full significance of the observed phenomena. For example, from the year 1840 onwards, Theodor Hartig (1840–1851) had noted and put on record the webs of hyphae surrounding the tips of the fine absorbing roots in Conifers, and had observed also the network of mycelium between the cells of the outer cortex of the roots. He was mistaken, however, in his interpretation of the real nature of these structures, which he regarded as purely parasitic. Fabre (1855–1856) also had observed filamentous threads and skeins in the root-cells of various Orchids without realising their true nature, which indeed was first clearly recognised by Schacht in 1854.

Describing the roots of *Neottia*, Prillieux (1856 *a*) says:

> Les cellules de la racine de *Neottia nidus avis* portent des nuclei d'une grosseur extraordinaire, et sur lesquels j'ai plusieurs fois distingué deux nucleoles;... Les cellules contiennent de la matière brunâtre renferment également des filaments enroulés sans ordre autour de la masse solide qui occupe le milieux de la cellule. Ces filaments sont creux; ce sont de véritables tubes dont on distingue avex certitude les parois.

From the year 1860 onwards there was a great quickening of interest in aspects of biology relating to nutrition, more especially in those concerned with differences in the nutrition of plants as compared with animals. Before that date, the nutritive processes of plants and animals were believed to differ essentially,—the plant building up complex bodies from simple substances, the animal

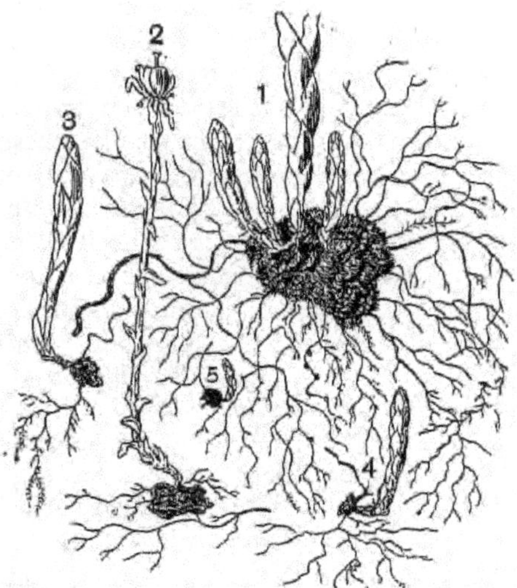

Fig. 1. Illustrations of the mode of growth of *Monotropa hypopitys*. (1) Base of a mature plant, 14 inches high, and three young unexpanded plants, growing from their radical parasitical knob. (2) Smaller plant in seed. (3), (4) and (5) Young plants growing from radical knobs. (Figures and description from Lees, *The Phytologist*, 1844.)

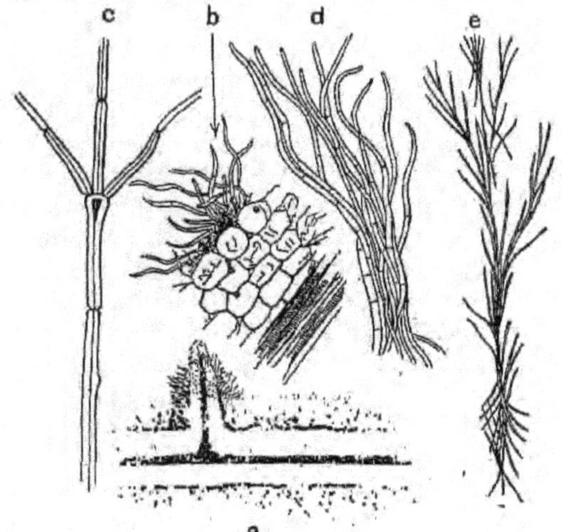

Fig. 2. (*a*) A section of the root of *Monotropa*, showing the internal continuation of the vascular tissue of the spongiole or fibril. (*b*) A portion of the same highly magnified, to show the connection of the flocci. (*c*) *Epiphagos luxfordii*, highly magnified. (*d*) and (*e*) The same, showing its adnate and fasiculate habit. (Figure and description from Rylands, *The Phytologist*, 1844.)

breaking these down again. The researches of Schultz (1861–1863) established the fundamental similarity of plant and animal protoplasm, thus leading to a search for the real causes for the differences believed to exist in the two groups in respect to nutrition. Following quickly upon this came a realisation of the specific property of green plants in relation to the manufacture from raw materials of substances suitable to be used as food materials by animals, with which was associated a newly awakened interest in exceptional modes of life and nutrition, e.g. in the behaviour of insectiverous plants, and in the presence of fungi and bacteria in the animal body.

Owing to their lack of chlorophyll and inability to utilise carbon dioxide, the Fungi stood out as possessing a mode of life fundamentally different from that of green plants. Observations by Pasteur in 1860 and 1862, by Nägeli in 1879 and 1883, and by Reinke in 1883 all showed that the former could use a great variety of carbon compounds as food materials. In 1874 so noted a botanist as Hooker, speaking of the Fungi, observed:—"these plants seem to invert the order of Nature and to draw their nutriment, in part at least, from the animal kingdom, which it is held to be the function of the vegetable kingdom to sustain" (J. Hooker, 1874).

The increased interest in problems of nutrition and the rapid advance of knowledge in this branch of biology were responsible also for the attention attracted to symbiotic phenomena generally from 1860 onwards. The true nature of the Lichen thallus, foreshadowed by de Bary in 1866, had been confirmed by Schwendener (1867, 1870) and established by the experimental researches of Stahl (1877), Reess (1879) and others. In 1879 the word *symbiosis* was coined by de Bary to cover all cases involving the living together of dissimilar organisms however loose the association. In his own words:—"...eine Betrachtung der Erscheinungen des Zusammenlebens ungleichnamiger Organismen, der Symbiose, wie man kurz und allgemein sagen kann...." Thus defined, the term includes all cases of parasitism, and was clearly so intended and used by de Bary himself:—"Die bekannteste und exquisiteste Erscheinung der Symbiose ist der vollständige Parasitismus" (de Bary, 1879). The above extracts are taken from an address delivered by de Bary to a meeting of a German natural history society at Cassel in 1879. It was published as a separate not always easily accessible for reference, and preceded by a few years only the general recognition of mycorrhiza as a typical example of the phenomena included by de Bary under the term symbiosis.

CHAPTER II

THE SECOND PERIOD

The Second Period: 1880–1900—Kamienski—Frank: recognition of mycorrhiza as a morphological entity—Classification into ectotrophic and endotrophic forms—Theory of beneficial symbiosis—The relation of Truffles and other fungi to mycorrhiza—Hartig's theory of parasitism.

THE SECOND PERIOD: 1880–1900 circa.

The gradual recognition of root infection as a regular and extremely widespread phenomenon among vascular plants began during the first years of this period. Simultaneously, interest was awakened in its physiological significance and experimental research on the subject was undertaken in more than one of the continental schools of botany.

Many different circumstances contributed to this result. There was already in existence a considerable body of sporadic observations recording the presence of mycelium in and upon roots. Of these a large proportion had been made upon species which showed marked peculiarities of structure and habit, and to which the attention of botanists had long been directed in respect to their anomalous mode of nutrition. For example, the non-chlorophyllous orchids *Neottia nidus avis*, *Corallorrhiza* and *Epipogum* had already been worked at by Schleiden (1842), Schacht (1852, 1854*a*), Irmisch (1853), Prillieux (1856), Drude (1873) and Reinke (1873). The controversy among English botanists concerning *Monotropa* has been mentioned (p. 4) and will be more fully discussed below.

Contemporary observers could hardly fail to relate the invariable presence of mycelium in these remarkable plants with the anomalous mode of nutrition involved by their lack of chlorophyll, especially at a time when the attention of botanists was focussed upon problems of plant nutrition. The development of abnormal roots by certain trees and the presence of mycelium believed to be that of a parasitic fungus upon such roots had also attracted the attention of more than one observer.

In another field of work the discovery of the dual character of the Lichen thallus and the attention directed to symbiotic phenomena by de Bary had stimulated interest in relationships between green and non-green plants. The root tubercles of Leguminosae and those of *Alnus*, *Eleagnus* and other trees challenged attention, as did also the causes of gall-formation in general. Furthermore, the work of Pasteur and his school had prepared the way for a study of pathogenic conditions in vascular plants and their causation by parasitic fungi. The cytological changes in plant cells subjected to invasion had also come under observation and were available for comparison with those in mycorrhiza tissues, while the improvement in technique and optical apparatus was reflected in the increased accuracy of microscope observations and also in their greater cytological significance. Evidently the time was ripe for a more general recognition of the prevalence of fungal infection in roots, the possible significance of this phenomenon in plant nutrition, and its relation to parasitic infection of the ordinary kind.

Doubtless many of these circumstances were operative in leading Kamienski to re-investigate *Monotropa hypopitys*, a species which, in spite of the attention lavished upon it by botanists, was still the subject of many conflicting statements in botanical literature. The extent and scope of this earlier work on *Monotropa* may be judged from the account already given and from the following brief review.

The anatomy had been investigated by Unger (1840), who recorded the absence of true vessels from the wood, an observation not confirmed either by Solms Laubach (1867–8) or by Drude (1873). A paper describing the presence of "einen eigenthümlichen Stoff" in *Monotropa hypopitys* was contributed by Reinsch (1852). The minute seeds early attracted attention; their structure was described by Muller (1848), who, however, mistook the combined endosperm and embryo of the seed tissues for the embryo alone. Solms Laubach (1874) and Koch (1882) recognised the distinction between embryo and endosperm, the structural details supplied by the latter being amply confirmed by Kamienski in his memoir published in the same year. The supposed parasitic habit of the plant had long been a subject of controversy. Some of the views current have been mentioned; those of other observers may be summarised as follows:—In 1840 Unger had placed *Monotropa* in his seventh order of parasites, a view confirmed by Brandt (1869), who included this species in the same class as the Orobanches. Chatin (1865) had described and figured the seed, and had published an account of seedling development. According to this observer the plant was a parasite in the young stages, penetrating the root of the host by means of its thread-like base: at a later stage of growth, the basal part of the seedling perished, the organic connection between parasite and host disappeared and the mature *Monotropa* plant was nourished entirely from the soil. Drude had reached a somewhat different conclusion. Having observed young seedlings of *Monotropa hypopitys* growing among Pine needles, he concluded that the plant was a saprophyte in the early stages, becoming parasitic in the mature condition. He described and figured *Monotropa glabra* as a parasite, the roots of which penetrated those of Beech and Pine and drew nourishment therefrom. As in the earlier controversy on the same subject among English botanists the last word lay with those who held a contrary view respecting the habit of the plant. Thus, in discussing *Monotropa uniflora*, W. Hooker threw doubt upon its parasitism, pointing out that plants could be raised sown upon humus, independently of any host[1].

Following upon a special study of the vegetative organs Schacht (1854*b*) was even more explicit, stating that *Monotropa* formed no organic connections whatever with a host and was not parasitic at any stage of its life; he concluded that, like *Neottia*, it can nourish itself on the decaying products of certain plants and for that reason is always found in their near neighbourhood. Solms Laubach (1862) could find no haustoria or other evidence for regarding the plant as a parasite, a view in which Duchartre (1846) concurred. From this brief survey, it is evident that the structure and habit of *Monotropa* still offered a promising field for investigation, especially in respect to the fungus investment of the roots. Advantage of this fact was taken by Kamienski, whose classical researches on the subject, although they ante-dated the bestowal of the name "mycorrhiza" upon a corresponding structure in the roots of Cupuliferae by Frank in 1885, must be regarded as the first milestone on the route leading to the present knowledge of the subject.

Kamienski published a preliminary paper in 1881, in which, after criticising some of the views expressed by earlier workers, he gave a detailed anatomical description of the vegetative organs of the plant and described the mycelial sheath present on all the roots. The conflicting nature of the earlier views were thus described:—"...les résultats donnés par les auteurs de ces différents travaux sont si peu concordants, qu'il est absolument impossible d'en conclure quelque chose de positif au sujet de la structure, du développement ou de la manière de vivre de cette plante...." His own observations confirmed the view that haustoria were not present,

[1] Plants of *Monotropa uniflora* were reported to have been raised at the Glasgow Botanic Garden on soil brought from Montreal, but it was not stated specifically that seeds were observed to germinate (see W. J. Hooker, 1825).

and led him to conclude that in respect to nutrition the plant was a non-chlorophyllous saprophyte living upon humus in the soil. The structures described and figured by Drude as haustorial connections in *Monotropa glabra* he interpreted as fungus-deformed roots of Conifers growing among those of the former plant and closely resembling them in respect to the presence of a fungal sheath. In his account of the mycelial investment of the root of *Monotropa hypopitys* he noted that it was present in all roots, forming in each a continuous sheath which thinned out over the extreme tip:—"Toutes les parties les plus actives de la racine sont recouvertes d'une couche épaisse et dense d'un mycélien qui ne permet pas aux racines d'avoir un contact direct avec le sol." With reference to the absorption of food material, Kamienski recognised that all soluble nutrients taken in by the roots from the soil must pass through this fungal zone, and he raised the question of its beneficial or other effect upon the plant. A satisfactory answer to this question, he admitted, could be supplied only by the extension of his researches to include the raising of plants free from fungal infection by means of seed cultures.

This preliminary paper was followed by a fuller account including a critical survey of earlier work and a full and accurate description of the morphology and anatomy of the vegetative organs illustrated by excellent drawings (Kamienski 1882). As regards the habit, Kamienski confirmed his earlier view that no evidence whatever existed that the plant was a parasite. This memoir may therefore be regarded as closing the long controversy upon this matter, although it left unexplained the exact nature of the *saprophytic* mode of nutrition assumed to exist, and contributed no certain information upon the part—if any—played by the root fungus in the nutrition of the vascular plant.

Experimental work in seed germination undertaken by Kamienski gave no results. Seeds sown on many different substrata, e.g. humus, peaty soil, and manure, all failed to germinate, and the conclusions expressed in this paper depend, therefore, upon facts of observation only. Historically, they are of interest, inasmuch as the relation between root fungus and vascular plant was, for the first time, clearly defined as differing essentially from that existing in cases of ordinary parasitism, and there can be no doubt that the rights of priority in this matter belong to Kamienski rather than to Frank.

In relation to the subject of root infection generally, the paper has somewhat special interest because the author noted and carefully described the fine, fungus-infected roots of the trees under which *Monotropa* commonly grows, especially those of Beech, of which he supplied a figure. He commented upon the arrest of growth and increased branching exhibited by these roots, and also upon the absence from each of a typical root-cap and the invariable presence of a continuous sheath of interwoven hyphae, branches from which penetrated between the cells of the epidermis and formed a network separating the cortical cells one from another. In Conifers generally, and especially in Pine, the multiplication of lateral branches was very conspicuous and was noted as exactly resembling the typical dichotomy found in Lycopodiaceae.

More especially was Kamienski impressed by the similarity of habit, texture, and structure in the roots of Beech and those of *Monotropa* with which they were often closely associated. He believed, doubtless correctly, that to this fact was due Drude's error of observation respecting the existence of haustoria in the latter.

Similar roots showing abnormal structure had already been noted and described for several different trees by Janczewski in 1874 and by Bruchmann in *Pinus sylvestris* in the same year. It was generally held that the condition of such roots was due to parasitic invasion, and Reess (1880) believed the fungus present in *Pinus* to be *Elaphomyces granulatus*, the False Truffle or Hirschtruffle, the fruits of which he had found in quantity on roots of this tree. Boudier (1876) had already published a paper suggesting the probability of a parasitic habit in certain species of the genus *Elaphomyces*. Kamienski accepted this view, and although positive evidence for the identity of the root fungi of *Monotropa* with the genus *Elaphomyces* was lacking, believed that the same or a similar fungus was parasitic upon the distorted roots of Pine and Beech found interlaced among those of *Monotropa*. In a paper published in 1886, he stated his conviction that, in the mycorrhizas of trees, the roots were subject to attack by parasitic fungi without deriving any benefit from the latter; the mycorrhiza of *Monotropa*, on the contrary, he held to be a symbiotic association in which nutriment was conveyed to its vascular host by a non-parasitic fungus.

Lacking experimental evidence as to the behaviour of the vascular plant in the absence of its root fungus, Kamienski's views concerning nutrition belong to the region of speculation rather than to that of scientific fact. Inasmuch as the results of experimental research on this subject are not yet available, they are still of theoretical interest and may be criticised in the light of recent experiments on analogous cases. The hypothesis advanced to explain the nutritive relations in *Monotropa* was as follows. The plant grows in soil rich in humus and absorbs the whole of its food supply from the soil by means of the roots. Owing to the presence of a thick and continuous sheath of mycelium, the absorbing surface of the roots has no direct contact with the soil, and all soluble food materials entering the plant from that source must pass through the mycelial envelope. No evidence of parasitism on the part of the fungus was found and it was held that the hyphae grew upon the surface of the roots merely as on a convenient base offering a larger surface than the surrounding soil particles. In return for this hospitality the fungus was believed to provide nourishment for its host—"fournit au *Monotropa* de la nourriture." Physiologically, the mycelial mantle was assumed to function as the epidermis of the root, and the hyphae extending outwards from it as root hairs. It was further concluded that the demands made upon the fungus by the vascular plant could not be excessive since the former continued to use the roots rather than the soil as a substratum favourable to growth. With regard to the nature of the nutritive materials, it was held that both fungus and vascular plant could utilise the organic compounds present in humus, but Kamienski also put forward a view—believed by him to be strongly supported by the observational facts,—namely, that the mycelium on the roots of *Monotropa* was continuous with that growing *parasitically* in the roots of the neighbouring trees. This hypothesis rested upon the assumption that the mycelium upon the roots of *Monotropa* was identical and continuous with that in and upon the roots of Pine and Beech. Its acceptance or otherwise did not affect his general view as to the reciprocal relation existing between the two symbionts in the case of the former plant. "Cette relation étrange entre le champignon et le *Monotropa* n'est pas un fait unique et isolé dans la nature. Nous pouvons le ranger avec d'autres faits semblables auxquels M. de Bary a donné la dénomination de 'symbiose.'"

The fundamental weakness of the view put forward by Kamienski—as was undoubtedly realised by himself—was the complete lack of experimental evidence showing the dependence or otherwise of the *Monotropa* plant upon its fungal partner. Could the vascular plant absorb the necessary organic matter from the soil humus without the intervention of its root fungus? There is a lack of precision in the statements bearing on this that has remained characteristic of the subject as presented in botanical text books ever since.

On the one hand, it is definitely stated that the roots have no direct contact with the soil, on the other hand, it is implied that the mode of nutrition of each of the partners is similar in respect to the utilisation and absorption of organic compounds in the soil. To accept the view that a fungus can utilise the humus constituents of soil and at the same time grow parasitically upon roots of various trees presents less difficulty to the modern botanist than it did to Kamienski. It is more difficult to make the assumptions demanded by his hypothesis that mycelium of one and the same species can simultaneously grow *saprophytically* in the soil, *parasitically* upon roots of Beech and *symbiotically* upon roots of *Monotropa*. Kamienski obviously inclined to the view that the vascular plant profited from the association, although he evaded the consequences involved by this hypothesis, i.e. that *Monotropa* is directly parasitic on the fungus and thus indirectly parasitic upon the roots of the surrounding trees.

In all modern botanical textbooks, *Monotropa* is classed as a saprophyte. Whether this is true in the strict meaning of the word, or in exactly in what sense it is to be understood, remains as uncertain now as when Kamienski published his paper in 1882. Nor can any positive statement yet be made regarding the possible identity of the root fungus or fungi of *Monotropa* with those present in what was afterwards called the ectotrophic mycorrhiza of Pine and Beech.

On the other hand, the identity of the fungi concerned in the formation of mycorrhiza in *Pinus* is no longer a matter of surmise (see Chap. VIII), although that of the forms present in Beech has not yet been established with certainty.

Kamienski's memoir on *Monotropa* has been considered in some detail because his work has been somewhat unfairly overshadowed in the later literature by that of Frank. There can be no doubt that the former first recognised and stated the existence of a reciprocal relation between flowering plant and fungus in *Monotropa*. This claim was emphatically put forward at the time by Woronin (1885 *a*), who, after the publication of Frank's first paper, concluded a somewhat fiery criticism as follows:—"Alle Prioritätsrechte in der Frage über die auf Wurzelsymbiose beruhende Ernährung gewisser Bäume durchunterirdische Pilz mussen demnach nicht Herrn B. Frank, sondern Herrn Fr. Kamienski zugeschrieben werden."

As a matter of fact, priority in this matter as a whole belongs neither to Kamienski nor to Frank, but to Pfeffer (1877), who ascribed to the Orchid fungi a physiological rôle analogous to that of root hairs. His views on the physiology of the relationship in Orchids were clearly stated as follows:

> Hier (bei *Neottia*) kann man in der That nicht umhin anzunehmen, dass eine Association vorliegt, aus welcher auch die bewirthende *Orchideen* Nutzen zieht, indem sie von dem parasitisch und saprophytisch lebenden Pilze Nährstoffe empfängt, welche die Pilzfäden aus den Boden aufnahmen....Ich bin zwar überzeugt dass die *Orchideen* auch ohne solche Pilze bestehen können, indess ist damit nicht ausgeschlossen, dass die Pilzfäden, da wo sie vorhanden sind, ihrem Wirthe Nutzen bringen.

Although the justice of Frank's claim to priority is open to question, the importance of his work on the subject needs no emphasis. Indeed, the scanty space assigned to mycorrhiza in botanical text books is occupied chiefly with the results of his observations and with speculations upon their significance. His first paper in April 1885 was quickly followed by a number of others in which he extended and co-ordinated his earlier observations. Not only did he carry out extensive observations himself but his example stimulated many other workers to research in the same field, as is testified by the large literature contributed subsequently by his colleagues and students.

The investigations carried out by Frank were undertaken in the first instance at the request of the German State Forestry Department in connection with the proposed cultivation of truffles in Prussia. They had as a starting point the observed association of truffles with certain trees, especially with Beech, Hornbeam, and Oak, in conjunction with the researches of Reess (1880), who had described what he regarded as a parasitic infection of the roots of *Pinus sylvestris*, the mycelium responsible for which he believed to be that of *Elaphomyces granulatus*, fructifications of which were abundant among the same roots. Suspicion was thus aroused as to a possible parasitic relation of the mycelium of truffles with the living roots of trees, and Frank's original investigation was designed to test this possibility. As an immediate result of the work he reported the invariable occurrence of fungus infection in the roots of certain trees, especially members of Cupuliferae, such infection being, in his opinion, not parasitic but of definite service to the trees in relation to the absorption of water and mineral salts from the soil.

Frank regarded this regular association of root tissues and mycelium as a morphologically distinct organ for which he proposed the name *Pilzwurzel* or *Mycorrhiza*. He recorded the characteristic coralloid growth shown by the mycorrhizas of Cupuliferae and described their structure in detail. He found that mycorrhiza occurred whether truffles were present or not but did not on this account reject the view that *Elaphomyces* or some other member of Tuberaceae was the fungus species concerned, since it was not unusual for mycelium to remain sterile over long periods of growth and special conditions might be necessary for the production of fructifications.

He assumed that infection took place from the soil, and carried out experiments with water cultures in order to test the behaviour of young trees when grown without their root fungus. By this means, three year old seedlings of Oak were obtained showing a well-developed root system without fungus infection. If seedlings possessing mycorrhiza were transferred from soil to culture solutions, they showed a gradual reduction of infection in the newly formed roots. From such experiments Frank concluded that the root fungi could best perform their beneficent service to the trees when growing in soil, although no satisfactory evidence was produced that the growth of the trees concerned was deleteriously affected when they were absent from the rooting medium.

Frank emphasised the close analogy between the thallus of a Lichen and mycorrhiza, each with its green and non-green constituent, and concluded that a true symbiosis existed in roots of Cupuliferae. With regard to the distribution of mycorrhiza he found it to be invariable in Oak, Beech, and Hornbeam, occurring quite independently of soil or situation. He recorded it also in certain species of Willow, and among Conifers in Pine, Fir, and Silver Fir. On the other hand, he gave in the paper under review a long list of native shrubs and trees which did not form mycorrhiza, including among them Birch (*Betula alba*) and Lime (*Tilia europaea*), both of which were afterwards recognised as typical mycorrhiza trees. He also commented on the absence of mycorrhiza from the ground flora of Beech woods, e.g. from roots of *Mercurialis perennis, Anemone nemorosa*, and other species in which typical mycorrhiza of a rather different kind from that in Beech has since been described.

On grounds such as these the earlier papers of Frank are open to criticism, as are also, in the light of modern experimental research, his provisional conclusions drawn from experiments. In the history of mycorrhiza the paper published in 1885 possesses a special interest as the first generalised account of a phenomenon now known to be even more widespread than was believed by Frank, and also as that registering the origin of the term *mycorrhiza*. The energy displayed by Frank in collecting and correlating evidence of the wide distribution of root infection and the surprising character of the facts he brought to light, provide ample excuse for the inclusion of minor inaccuracies. At this stage of the enquiry it was perhaps inevitable that the author's zeal should tempt him to make the facts fit his hypothesis rather than the hypothesis fit the facts (Frank, 1885 *a*).

The outstanding results of Frank's preliminary investigations were, firstly, the recognition of root infection as a widespread phenomenon in trees, and the bestowal upon it of a distinctive name marking its existence as a morphological entity; and, secondly, the rejection of the accepted view of parasitic invasion of these roots, whether by Truffles or other soil fungi, and the substitution of his theory of a symbiotic relationship beneficial to the trees. The far-reaching character of this hypothesis was a direct incentive towards the collection and interpretation of new facts bearing on the subject.

Later in the same year Frank (1885 *b*) published another contribution to the subject, provoked doubtless by Woronin's claim of priority for Kamienski. While tacitly admitting the latter's claim to priority of observation of root infection, Frank quite justly pointed out that the root fungi were believed to be parasitic on trees and assumed priority for his own interpretation of the association as a phenomenon of mutualistic symbiosis. His conclusions as to the physiological relationship existing in Cupuliferae were concisely stated in this paper and may be briefly summarised as follows.

Mycorrhiza is a symbiotic relation to which probably all trees under certain conditions are subject. It is formed only on soils containing humus or abundant plant remains, and its formation waxes and wanes with the abundance or otherwise of these constituents in soil. The root fungi carry to the trees not only the necessary water and salts but also soluble organic material derived from the humus, thus lending a new significance to leaf-fall and the accumulation of humus in woodland soils, and reviving incidentally an old theory concerning the nutrition of green plants in a sense somewhat different from that in which it was stated originally. The assistance thus rendered to the trees possesses special importance when rapid growth makes heavy demands upon available food constituents in the soil, and is likewise of great significance to non-chlorophyllous species.

During the two years following, Frank extended his observations in many directions and learned that the regular association of fungi with the roots of vascular plants may show structural characters very different from those found in Cupuliferae, Conifers, and *Monotropa*, regarded hitherto by him as the normal type. Recognition of this led to the publication of another paper in which his well-known classification of mycorrhiza to *ectotrophic* and *endotrophic* forms was proposed. "Wenn wir alle diejenigen Formen bei denen der ernährende Pilze sich auswendig befindet als *ectotrophische* und diejenigen, wo er das Innere gewisser Wurzelzellen einnimmt als *endotrophische* bezeichen, so erhalten wir folgende Uebersicht" (Frank, 1887 *b*).

The descriptive terms endotrophic and ectotrophic have been in general botanical use since that time. The distinction between typical and extreme cases of the two kinds is sufficiently obvious. Owing partly to defective observations and partly no doubt to the great weight attaching to Frank's views on the subject, the essential characters of the distinction between the two types was overemphasised by some of his contemporaries and successors. The use of the terms in a somewhat rigid sense in botanical textbooks has doubtless delayed the recognition of forms intermediate in structure between the two extreme types. Recent work on the cytology of ectotrophic forms confirms the view that mycorrhiza showing the structural characters of both types is not uncommon and that the presence of intracellular mycelium in the cortical cells of ectotrophic mycorrhiza is relatively frequent—in short, that the difference is one of degree rather than of kind.

The review which accompanied the analysis and classification of types in this paper of Frank's is of interest as a summary of the observations available at that time and merits brief notice.

Ectotrophic Mycorrhiza.

The distinguishing features of the ectotrophic type described by Frank in Cupuliferae, Conifers and other trees were:—(1) the invariable presence of mycorrhiza throughout the life of individual trees from the first year of growth onwards, and its distribution upon the actively absorbing regions of the root system; (2) the coralloid growth exhibited by infected roots or root systems; (3) the absence of root hairs and reduction of the root caps to a few cells only; (4) the presence of a complete investment or mantle of fungal hyphae covering the whole of the younger parts of the roots including the apices; (5) the existence within the root tissues of a continuous network of hyphae separating the individual cells of the epidermal and cortical layers, the fine hyphae composing the network being continuous with those in the external mantle and not infrequently penetrating the epidermal cells; (6) the absence of intracellular mycelium from the cells of the cortical region.

Frank recognised the variation that existed in the structure of the mantle in regard to thickness, coloration of hyphae, and nature of the surface, e.g. whether smooth or covered with projecting hyphae extending into the surrounding soil. He considered that such differences were largely specific to individual types of mycorrhiza and did not depend upon age or degree of development. Cases were described differing remarkably from what was regarded as the normal type. For example, an anomalous long-branched—i.e. not coralloid—mycorrhiza on Beech from Hanover with a very thick mantle and "pseudo root hairs" was described and figured. The outgrowing strands of hyphae from the surface of this mycorrhiza resembled root hairs both in appearance and in their relation with the soil particles and were believed to function in a manner quite analogous to these organs (Fig. 3). Another peculiar form was recorded upon roots of *Pinus Pinaster* from Capetown. This mycorrhiza, described as resembling a "fox's brush," consisted of a main axis, thickly beset with hair-like threads of approximately the same length. Microscopic examination showed the latter to be short, fine lateral roots, each of which was a typical ectotrophic mycorrhiza (Fig. 4*a*, *b*, *c*). This anomalous structure was not found in European material of *Pinus pinaster*.

Endotrophic Mycorrhiza.

Under this heading, Frank described the two types of mycorrhiza characteristic of Orchidaceae and Ericaceae respectively. In view of the fact that considerable space will be devoted to a consideration of these groups in later sections of the present work a very brief summary of the facts noted by Frank will suffice.

The condition of the roots in Orchids had already attracted the attention of a series of observers from the time of Schleiden onwards; of these, a contemporary, Wahrlich (1886), had published a paper dealing especially with the root fungi of a number of Orchid species. The general characters of Orchid mycorrhiza were therefore fairly well known; among the features specially noted by Frank were:—(1) the non-parasitic character of the fungus invasion of the roots and its importance in relation to the nutrition of the Orchid plants; (2) the disposition of the infected tissues in such a position that substances absorbed from the soil must pass through them on entering the plant, and (3) the invariable presence and high

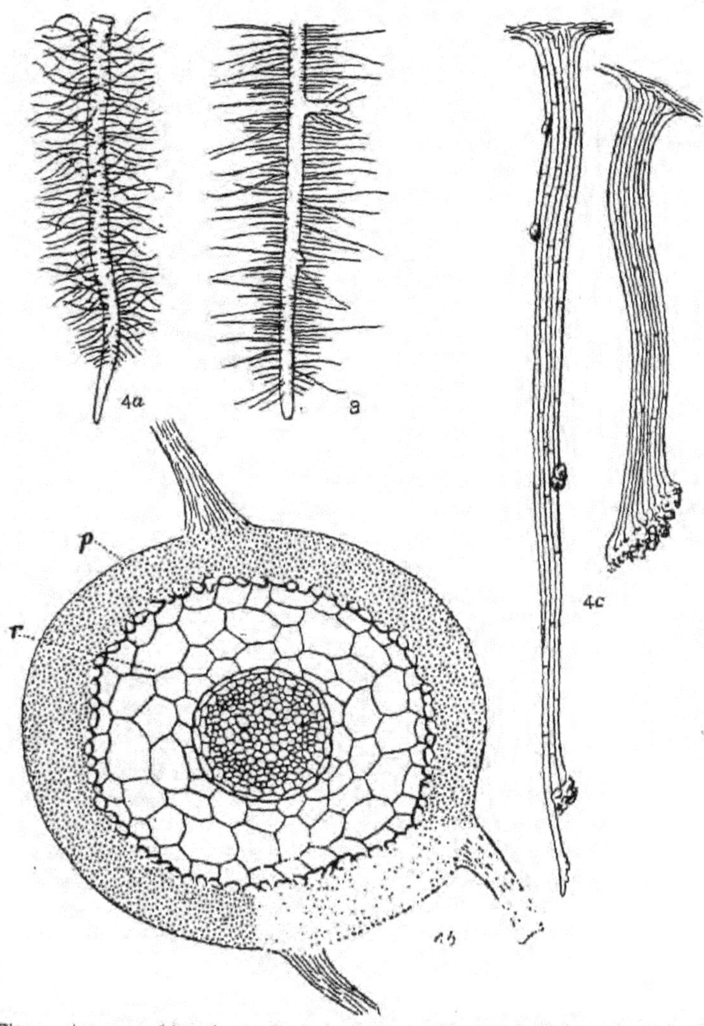

Fig. 3. A mycorrhiza from *Fagus sylvatica* with root-hair-like strands of mycelium. (From Frank, 1887.)

Fig. 4 *a*. Root of *Pinus pinaster* from S. Africa with root-hair-like mycorrhiza. (From Frank, 1887.)

Fig. 4 *b*. Transverse section of the mycorrhiza shown in Fig. 2: *p*, thick fungus mantel with outgoing strands of mycelium; *r*, cortex of root. (From Frank, 1887.)

Fig. 4 *c*. Two mycelial strands from the same mycorrhiza; to show the intimate relation with soil humus. (From Frank, *Ber. d. d. bot. Gesell.* 1887.)

degree of development of mycorrhiza in the non-chlorophyllous Orchids.

In respect to Ericaceae, external infection of the roots of *Andromeda polifolia* had been noted by Frank in 1885, and Thomas (1885) had recorded root infection in Bilberry and inaccurately described it as of similar type to that found in *Monotropa* and members of the Cupuliferae. Subsequently, Frank examined the roots of a number of other members of the family and contributed the first general account of mycorrhiza in the group in the paper under review.

Among the species mentioned by him are *Andromeda polifolia, Ledum palustre, Calluna vulgaris, Rhododendron ponticum, Azalea indica*, and a number of species of the genus *Vaccinium*, e.g. *Vaccinium oxycoccus, Vaccinium myrtillus, Vaccinium vitis idaea* and *Vaccinium macrocarpum*, all of which showed regular and characteristic infection of similar type. The general appearance of mycorrhiza in the fine roots of Ericaceae was admirably described by Frank. Attention was directed to the fact that in Ericaceae, as in Cupuliferae, the formation of mycorrhiza was independent of soil conditions *other than the presence of abundant humus*; e.g. the presence of a thin surface layer rich in humus was noted on sandy heaths carrying ericaceous vegetation. Among outlying members of the group, *Pyrola* was recorded as free from root infection and *Monotropa* noted as differing in the possession of mycorrhiza of the ectotrophic type.

Although not a member of Ericaceae, *Empetrum nigrum* was also included in this account of mycorrhiza in heath and moorland plants as a species possessing root structure and mycorrhiza resembling that found in the Heath family, with various members of which it is frequently associated in the field. With the exception of *Empetrum* and Ericaceae, mycorrhiza was not recognised by Frank in other members of moor and heath vegetation in North Germany; for example he recorded it as absent from the roots of *Aspidium thelypteris, Carex vulgaris, Drosera rotundifolia, Viola palustris, Menyanthes trifoliata, Comarum palustre, Tormentilla erecta*, and *Trifolium repens*.

Observations on mycorrhiza in Germany were confirmed by examination of material sent from other countries and by the accounts of independent observers. Thus, records from S. Africa for Cupuliferae, Conifers, and Heaths, from Australia for *Fagus Cunninghami*, and from New Holland for Epacridaceae, gave evidence of its wide distribution, and confirmed the view expressed by Frank as to the invariable occurrence of mycorrhiza in certain species however wide their geographical range. Möller confirmed Frank's observations on Beech for Danish material and Ascherson recorded the presence of mycorrhiza in roots of Alpine species of Willow (*Salix retusa* and *Salix reticulata*) up to 3000 metres.

The later publications of Frank (1888–1894) are concerned especially with the elaboration of his theory of nutrition in mycorrhiza plants, and the accumulation of experimental evidence in support of it. In their more general aspects these views have become an integral part of the literature of the subject. The observations and experiments on which they were founded are not so well known to modern botanists; in view of the fact that almost 40 years have elapsed since their publication and that they serve as a convenient starting point for a survey of recent work on the subject in a later section of the present review, the following brief recapitulation may serve a useful purpose.

Frank believed that the existence of the appropriate root fungi in soil—and hence the formation of mycorrhiza by trees—was conditioned by the presence in woodland soils of humus derived from fallen leaves and other organic residues of the trees growing in such soils. This belief he tested experimentally by comparative observations on Beech seedlings raised upon soils with and without humus, or transferred from their natural habitat in woodland soils to pots of sand. In general, the results of such experiments were interpreted as favouring the view that the root fungi were present in woodland soils and depended upon the humus present for their nutrition.

That humus from other sources was not necessarily effective for the purpose was deduced from experimental cultures; thus, in humus soils other than those of woodland origin, Beech seedlings were observed to grow unsatisfactorily, and failed to develop mycorrhiza. Similar conclusions followed from transplanting young Beeches from natural woodland soils to others lacking humus. The transplantations were successful, the seedlings established themselves afresh, but the new roots present at the end of two years were almost entirely deprived of mycorrhiza.

It was concluded by Frank that the root fungi of trees tended to disappear when humus of the appropriate kind was absent, i.e. that their proper nutrition depended upon an adequate supply of this soil constituent. In his own words:—"die Mykorhizapilzen nicht in den lebenden Pflanzenwurzel ihre Lebensbedingungen finden, sondern vielmehr in gewissen Beschaffenheiten des Bodens und dass es vor allen Dingen der Baumhumus ist, von welchem die Anwesenheit dieser Pilze und der von ihnen gebildeten Mykorhizen abhängt."

His further conclusions as to the relation of the root fungi of trees to the nutrition of their hosts were based mainly upon the structure of ectotrophic mycorrhizas, the correctness of the deductions being tested experimentally by sand and water cultures of Oak and Beech. The general character of these conclusions is familiar to all botanists and may be summarised briefly as follows.

The fungus mantle with its outwardly extending hyphae acts as the absorbing system of the tree, through which water and mineral salts must pass on their way to the vascular strands. The hyphae of the inner part of the mantle and those forming the internal network are continuous with those outside and are so placed that osmotic exchanges can take place with the epidermal and cortical cells with which they are in contact.

The existence of a similar type of mycorrhiza in *Monotropa*, a non-chlorophyllous species, in conjunction with the results of experiments designed to test the effect upon nutrition of withholding a supply of humus from Beech seedlings, led Frank to include the formation and intake of suitable compounds of carbon and nitrogen from humus among the services rendered by the root fungi to their hosts. Experiments were carried out to obtain evidence in support of this view. Thus, it was found that Beech seedlings germinated in sand or water cultures and, supplied with the necessary inorganic salts, did not form mycorrhiza, grew badly and died after one year's growth. Similarly, when parallel cultures of Beech were grown, using woodland soil untreated and the same soil sterilised at 100° C., all the seedlings planted in the former developed typical mycorrhiza and grew vigorously, while those raised in sterilised soil formed no mycorrhiza, died off one by one, and had all succumbed at the end of three years.

The experimental evidence was believed to prove that both the Beech and its specific root fungi required a supply of humus for their proper nutrition and could make but scanty growth if this was lacking. Deprived of mycorrhiza, Beech seedlings suffered inhibition of growth and eventually died because they were unable to utilise the food supplies locked up in humus. In other words, the trees depended upon their fungus symbionts for the re-conversion of their cast off residues into active plant substance. In a subsequent paper, similar experimental results were recorded for Pine (*P. sylvestris*), using comparative seed cultures in untreated soil from a Pine forest and similar soil steam-sterilised for one hour at 100° C. The seedlings in the untreated soil behaved as in seed-beds under natural conditions; those in the sterilised soil made poor growth, showed reduced size and discoloration of the needles and at the end of two years were in a moribund condition. The former developed typical mycorrhiza; the condition of the latter in respect to this is described as follows:—"die mikroskopische Prüfung der Wurzeln ergab aber keine Spur von Verpilzung, die Wurzeln kurze Wurzelhaare gebilden."

Since it had been shown previously that plants other than those with mycorrhiza grew well in sterilised soil, it was concluded that the roots of Conifers, like those of members of Cupuliferae, no longer function like the roots of ordinary plants but depend for their proper nutrition upon the presence of their appropriate root fungi.

Inasmuch as this was the case in trees and in *Monotropa*, these plants must be regarded as *parasites upon saprophytic fungi*. Thus:—"Insofern also der Baum seine Nahrung unmittelbar aus der Hand des Pilzes und erst durch dessen Vermittelung aus dem Erdboden empfängt musste man vielmehr die Waldbäume und die in den Wurzeln sich dessen gleich verhaltende *Monotropa hypopitys* als Parasiten auf saprophyten Pilzen betrachten." No ordinary parasitism this, but "eine mutualistisches Verhältniss, eine wirkliche Symbiose"!

These views were not endorsed by Henschel (1887) for the mycorrhiza of Conifers. On the contrary, the observations made by this worker led him to the view—"dass der Einfluss diesen Symbioten auf der Entwicklung der jungen Fichtenpflanze als ein absolut schädlichen angesehen werden müsse."

The belief that woodland humus in particular was an important source of nitrogenous food material was greatly strengthened by independent observations on the nitrate content of trees and other mycorrhiza plants on various soils. Frank and Molisch had previously shown that a direct relation existed between deficiency of nitrates in soils and a similar deficiency in the tissues of plants growing upon them, from which it was reasonably inferred that vascular plants were unable to convert either free nitrogen or compounds of nitrogen into nitrates. Moreover, Frank (1887 *a*) had recorded the absence of nitrates from the aerial parts of various trees and their absence, or very scanty appearance, in the absorbing roots, while the tissues of *Neottia* were found to give no reaction for nitrates although such is usually well-marked in herbaceous plants.

An additional link in the chain of evidence was provided by

Ebermayer (1888), who, in a contemporary paper entitled "Warum enthalten die Waldbäume kein Nitrate?" recorded the almost complete absence of nitrates from more than one hundred samples of woodland and moorland soils.

All these facts led independently to the conclusion that plants growing upon soils deficient in nitrates must obtain their supplies of nitrogen either as ammonium salts or as organic compounds of this element. They pointed also to the possible significance of the root fungi in this connection and, assuming likewise the importance of carbon compounds other than carbon dioxide for non-chlorophyllous species such as *Neottia* and *Monotropa*, justified the conclusions expressed by Frank in the following passage:—"bei den chlorophyllhaltigen Pflanzen die Mykorrhizapilze hauptsächlich die Erschliessung des Humusstickstoffes bewirken." His own experiments led him to believe that certain plants, e.g. Oat and Lupin, which were believed free from fungus infection, could directly utilise the organic compounds present in soil. It was pointed out also that potassium and other essential elements were locked up in the organic residues of humus and might be rendered accessible to vascular plants by the metabolic activities of their root fungi.

The theory of nutrition outlined above was extended by Frank to cover cases of endotrophic mycorrhiza, the number of which had been greatly extended by the researches of his pupil Schlicht (1889) and by those of Wahrlich (1886) on the Orchids. The failure of many of these plants to produce root hairs was noted, as was also the position of the infected cells in a zone of tissue between the absorbing and conducting regions of the root. Owing to the scanty development of mycelium *outside* the roots, it was impossible to relate the root fungi directly to absorption as in ectotrophic mycorrhizas; moreover it was uncertain to what extent the intracellular mycelium drew upon the root cells for nutritive materials. On the other hand, Frank had recently studied the mycorrhiza cells in certain Orchids—in particular in *Orchis latifolia*—and had been greatly impressed by the extensive digestion of mycelium in these cells and the subsequent removal of the products of digestion as deduced from staining reactions. Here, indeed, was a new and quite unique kind of symbiotic relation between vascular plant and fungus! He extended these researches on Orchids to the fine roots of *Ledum palustre* and *Empetrum nigrum*, which showed similar staining reactions with aniline blue and were described as quite analogous to the Orchids in this respect.

Frank therefore included all the known cases of endotrophic mycorrhiza as typical examples of mutualistic symbiosis, although he deplored the circumstance that experimental proof of the symbiotic relation, in the form of evidence as to the behaviour of infected and uninfected plants in comparative cultures, was still lacking both for Orchids and Heaths.

The evidence of intracellular digestion found in Orchids was compared with that exhibited by insectivorous plants, and the strength of Frank's convictions as to the beneficient rôle of endotrophic mycorrhiza is evidenced by the close analogy that he postulated between the two classes of plants. "Denn der Pilz ist hier gleichsam in den Wurzelzellen gegangen, wo er als sichere Beute zuletzt von der Pflanze aufgezehrt wird" (Frank, 1892 *b*).

Although he fully realised and admitted the importance of comparative cultures with and without fungus infection, the experimental evidence offered by him on this fundamental aspect of the subject is unsatisfactory in the case of ectotrophic mycorrhiza, and lacking altogether for the endotrophic forms, with which no experimental work was attempted. In a later chapter of the present review it will be of interest to compare the opinions of Frank, based largely on theoretical considerations, with the results of recent researches, for which pure cultures of certain trees and of their root fungi have been available.

Cogent as are many of the arguments based upon the structure and cytology of mycorrhiza and ably as the evidence is marshalled in these papers, it is difficult to avoid the imputation of special pleading to the part of Frank's work relating to experimental observations. Some, at least, of his results are open to interpretations other than those offered, and the experimental evidence is in itself too slight to carry the imposing edifice raised upon it.

His interpretation of the facts was not always acceptable to contemporary botanists; in particular, his views invited criticism from students of the effects of parasitic invasion by fungi, at that time a new branch of botanical science. In his first paper Frank had referred to the work of Gibelli (1883) on a disease of the Chestnut in Italy, criticising the opinions of the latter with regard to the relation of parasitic fungi believed to be responsible for the pathogenic conditions with the hyphae of the mycelial envelope likewise present on healthy roots. On behalf of Gibelli, Frank's criticisms on this matter were questioned by Penzig (1885) in a subsequent paper.

Hartig (1888 *a*) had described a disease of young Oaks attributed to parasitic attack by a fungus, *Rosellinia* (*Rhizoctonia*) *quercina*, but the condition of the roots showed no agreement with that of the mycorrhiza of Oak described by Frank.

Publication of the latter's theory of symbiotic nutrition in 1885 quickly provoked a criticism from Hartig (1886) and Grosglick (1886), who put forward the view that mycorrhiza in general is a pathological condition brought about by the attack of parasitic fungi. In reply, Frank emphasised the relative longevity of mycorrhizas as compared with roots subjected to ordinary parasitism, the active growth of the former subsequent to infection, and his experimental demonstration that in humus-free soil the fungi desert the roots instead of parasitising them. The arrest of growth in length and increased branching, noted by Hartig as symptoms of disease, were regarded by Frank as adaptations securing concentration of the absorbing system in those regions of the soil where humus was most abundant. In support of this view, he cited the well-known fact that a rich substratum promoted abundant branching of roots with relatively limited growth in length, whereas one relatively poor in nutrients led to scanty branching with greater elongation.

Hartig (1888 *b*) specially emphasised his own observations that young roots of trees were free from fungus infection during the period of maximum absorption from June to September. As they became infected from older roots this functional activity diminished and might cease altogether until a fresh supply of young roots developed in the following spring. Kamienski (1886) also disagreed with Frank's theory of beneficial symbiosis in the case of trees, pointing out that infected roots were less common than reported and that the tissues were often pathological in appearance. He re-stated his own view that the relation in trees was one of parasitism on the part of the fungus, while in *Monotropa* it was one of mutual symbiosis resulting in a supply of nutrient to the host plant. He claimed that the chief merit of Frank's work consisted in drawing the attention of botanists to a phenomenon which invited research and was worthy of greater attention.

The challenge issued by Woronin (1885 *a*) in respect to the priority of Kamienski's observations has already been noted. In the same paper this author states that the regular root infection described by Frank had been independently observed by himself in roots of various members of Cupuliferae, Salicaceae and Coniferae, while working on certain edible species of *Boletus*. Woronin did not challenge Frank's interpretation of the facts, beyond stating his own conviction that the relation was one of parasitism on the part of the root fungi. He pointed out also that in Finland, where many of his own observations had been made, the mycelium present in roots was not likely to be that of Truffle fungi since the latter were not found in that country.

Papers dealing with the same subject, i.e. the parasitism of Truffle fungi and their relation to mycorrhiza, were also published by Müller (1886) and by Mattirolo (1887); the latter reached the provisional conclusion that a parasitic mycelial investment of the roots of certain trees—"rizomorfa parassite delle radici"—in all respects similar to those described by Frank under the name of mycorrhiza, could, under conditions not yet fully ascertained, give rise to the reproductive bodies of *Tuber excavatum*.

The identity of the root fungi was still a subject of controversy. Tulasne (1851) had noted that *Elaphomyces* formed a coating upon roots of Pine regarded by him as of parasitic nature. Later (1862), he modified this view and expressed the opinion that the two constituents derived mutual benefit from the association.

Reference has been made to the work of Reess (1880) who had suggested that the mycelium in mycorrhiza was that of *Elaphomyces granulatus*, a fungus believed to be related to the Truffles. His further researches (Reess 1885, 1887) had demonstrated the striking similarity of the mycelium in roots of *Monotropa*, Conifers and members of Cupuliferae, but provided no positive proof of its identity with that of *Elaphomyces*; moreover, they introduced the difficulty of knowing whether one or a number of fungal species were concerned in the formation of mycorrhiza. In Pine, for example, upon the roots of which spore bodies of this genus had been found, there was evidence of the presence of another type of mycelium resembling that characteristic of Basidiomycetes. It is clear from his later papers that Reess had an open mind on this aspect of the subject, and, without rejecting the opinions expressed by Kamienski and Frank with respect to nutrition, believed that further evidence was required. Thus, as arguments against the symbiotic view he cited, (1) the absence of mycelium from the roots of many woody plants, (2) the occasional abundance of root hairs in mycorrhiza plants, e.g. in Pine, and finally, (3) the numerical relation between infected and uninfected roots in Pine. While respecting Frank's views on the subject it was clear that he did not agree with them.

Referring to the paper by Woronin on this subject, Lecomte (1887), recording his own observations in the Vosges, stated that conidia and perithecia observed by him on roots of Hazel were produced by the mycorrhizal fungus specific to the tree; the reproductive structures observed were believed to indicate affinity with members of Perisporiaceae. In the same paper reference was made to similar observations by an Italian mycologist, Cornu, in conjunction with Gibelli (*loc. cit.*), who also held that the mycelium in roots of Cupuliferae, Conifers, etc., belonged to species of *Elaphomyces* or *Hymenogaster*. That other aspects of the subject also attracted attention is evidenced by the observations of P. E. Müller (1903) on the nature of humus and the distribution of fungi growing in it.

Noack (1889) investigated the relationship between the mycelium of various Hymenomycetes and Gasteromycetes and the mycorrhiza of certain trees. For example, he observed that coralloid roots of *Pinus sylvestris* were almost invariably present in the humus below the sporophores of *Geaster fimbriatus*. By similar observations on fruit bodies Noack was led to include the following fungi as mycorrhiza formers: *Geaster fimbriatus* and *Geaster fornicatus* on *Pinus sylvestris* and *Abies excelsa*; *Agaricus* (*Tricholoma*) *russula* on Beech, forming the rose-pink mycorrhiza described by Frank; *Agaricus terreus* on Pine and Beech, *Lactarius piperatus* on Beech and Oak; *Cortinarius* sp. on Fir, Beech and Oak. He noted also that a rose-coloured mycorrhiza on roots of *Pinus sylvestris* was clearly associated with fruit bodies of *Elaphomyces granulatus*. Unfortunately, Noack's conclusions were based almost entirely upon microscopic examination of roots and the propinquity of fruit bodies; attempts to reproduce mycorrhiza from spores of *Geaster* sp. in fungus-free seedlings of Fir,—"die gewiss fur die ganze Frage von grosster Wichtigkeit sind,"—gave only one case of doubtful infection.

Mollberg (1884) and Wahrlich (1886) had made further contributions to the already extensive literature dealing with the mycorrhiza of Orchids. The latter writer does not refer to Frank's work and throughout alludes to the root fungi of the Orchids as parasites (Wurzelparasite). His researches sought to determine the real nature of the yellowish slime masses in the root cells, to learn whether a single fungus species was responsible for infection throughout the group, and to observe the mode of fructification of the endophytes with a view to their proper classification.

Examination of a number of European Orchids and of some 500 exotic species showed that the roots of all were more or less heavily infected by mycelium of the same general type. Microchemical investigation of the intracellular slime masses described by previous observers confirmed the view that they consisted of tightly wound hyphae embedded in an oily or resinous material believed by Wahrlich to be produced by the mycelium. He was not so successful in his efforts to cultivate and identify the endophytes. The technique adopted was crude, and mycelium from a number of Orchid species when cultivated outside the plants gave rise to spores of the *Fusiformium* type described earlier by Reissek and by Schacht. The fungi isolated from several species of *Vanda* produced perithecia, thus leading Wahrlich to place them in the genus *Nectria*. In his view, the general similarity of structure in the mycelium associated with roots pointed to the inclusion of all the root fungi in one generic group, minor differences in the structure of the mycelium and spores suggesting that this comprised more than one species.

Attempts to cause re-infection of young aerial roots by means of spores and thus establish the identity of the forms isolated were unsuccessful, in spite of which Wahrlich committed himself to the following conclusions:

> Fruchtificationsorgane des Wurzelparasiten der *Orchideen* sind *Fusiformium*sporen, Megalosporen und die bei *V. suavis* und *V. tricolor* beobachteten Perithecien mit Ascosporen....Nach seinen Fructificationsorganen ist der Pilz als ein *Pyrenomycet* zu bezeichnen. ...Auf Grund dieser Merkmale sind die in *Vanda* Wurzeln parasitirenden Pilze als *Nectria*arten zu bezeichnen....

This paper by Wahrlich is frequently quoted, although it contributed little new to the contemporary knowledge of Orchid mycorrhiza. The major part is devoted to an account of the attempts made by the author to isolate the root fungi from various Orchids and to cultivate them outside the plant; the conclusions reached were entirely uncritical and were unsupported by any experimental proof of the identity of the fungi isolated with the true endophytes.

Schlicht (1889), working in Frank's laboratory, published an account of his researches on endotrophic mycorrhiza in herbaceous plants. He described in detail that found in *Paris quadrifolia*, *Ranunculus acris*, *Holcus lanatus* and *Leontodon autumnalis*, noting its restriction to the actively absorbing roots and the position of the infected tissue between the absorbing and conducting regions. He observed, moreover, that mycorrhizas not uncommonly persisted as main roots, and that the presence of intracellular mycelium offered no impediment to growth, thus controverting the conclusions of Hartig with respect to parasitism.

The possession of this "neuen Mykorhizen" was recorded by

Schlicht for some 112 herbaceous species distributed over about 45 families of Angiosperms, and it was concluded "dass die Symbiose zwischen Wurzeln und Pilzen eine ungeahnte Verbreitung über unsere Flora hat." Observations such as that recording complete absence of root mycelium from certain groups, e.g. Gentianaceae, certainly required revision, as did also the list compiled by Schlicht of mycorrhiza-free species belonging to 22 families. Reviewing the published work of Kamienski, Reess and Frank on the subject, Schlicht pointed out that the relation involved parasitism on the part of either vascular plant or fungus, or a condition of mutualistic symbiosis. In view of his association with Frank, it is perhaps not surprising to find that his own view favoured the last alternative:—

dass die von Frank aufgestellten Thesen über das Wesen und die Bedeutung der Mykorhizen der Bäume durch die nunmehr erkannte Verbreitung der endotrophischen Mykorhiza eine noch viel weitere Ausdehnung erlangt haben, denn aus meinen Beobachtungen ergiebt sich das überraschende Resultat, dass auch die chlorophyllführenden Kräuter im Allgemeinen nicht ohne einen Wurzelpilz auftreten, und dass auch sie anscheinend nicht ohne diesen bestehen können, wenn ihnen nicht die aus dem Boden aufzunehmende Nahrung in einer anderen ihnen zusagenden Form geboten wird.

CHAPTER III

HUMUS PLANTS

Humus plants—Saprophytes and hemisaprophytes: Höveler; Johow; Groom; Janse; Hesselman; MacDougal—The cytology of intracellular digestion: Magnus; Shibata.

1890–1900[1].

The recognition of root infection as a regular and widespread phenomenon, and the attention attracted to the physiology of the relationship by the new theory of beneficial symbiosis, stimulated investigation in the subject during the closing years of the nineteenth century, more especially in relation to the study of the endotrophic type of mycorrhiza.

Some of the researches, e.g. those of Schlicht, mentioned above, were carried out in Frank's laboratory, and indeed many of those published during the earlier part of the period were obviously influenced by his point of view and directed to extend and confirm his conclusions. During the closing years of the nineteenth and the first five years of the twentieth century the cytology of the infected root cells specially attracted attention, and knowledge of various aspects of this part of the subject was extended by the researches of a number of English and continental botanists. Many of the observations published about this time were made upon Orchids which afforded exceptionally favourable material for the study of the progressive changes in the infected root cells. The following survey of the more important papers published during the concluding years of the last century will serve to indicate the quickening interest of botanists in the subject of mycorrhiza and its increased importance as a branch of botanical investigation.

Mention may first be made of the comprehensive review of work on root symbiosis and mycorrhiza published by Sarauw (1892). Numerous facts of historical interest are preserved in this paper, which included an exhaustive and accurate bibliography of the literature of root-infection, parasitic and otherwise.

About the same time appeared a paper by Höveler (1892) dealing with the soil humus as a source of food material for green plants.

[1] Frank's *Lehrbuch der Botanik* was published in 1892, and included a summary of his observations on mycorrhiza and his views respecting its significance.

The old view as to the importance of humus in this connection had long been discredited, but little was known as to the exact chemical nature of this important soil constituent. The subject had engaged the attention of a number of workers, and numerous publications had already appeared dealing with the origin and nature of the soil humus and the utilisation of organic compounds by green and non-green plants whether by direct absorption, or indirectly, through the agency of root fungi.

The researches of Frank (1888 *b, c*) and Koch (1887) on this subject have been mentioned. Among other contributors to the literature of the subject were Stutzer and Klingenberg (1882), Baumann (1886), Müller (1887, 1889), Déhérain (1889), Hoppe-Seyler (1889), Acton (1889), and Berthelot and André (1892).

Höveler devoted special attention to the presence or absence of root hairs and the incidence of fungus infection in humus plants, an aspect of the problem indicated by Frank's insistence on the absence of root hairs from plants with ectotrophic mycorrhiza, and their relative infrequence in those showing the endotrophic type of infection. He recorded abundant root hairs of a rudimentary type in Orchids generally, noting also the presence of endotrophic mycorrhiza in *Epipactis latifolia*, which, in common with *Listera ovata*, Frank had previously described as free from fungus infection. In general his observations lent support to the view that plants with abundant root hairs were free from root infection, although *Calla palustris* was noted as a species possessing both root hairs and mycorrhiza. Among humus plants producing root hairs but growing under field conditions favourable to the formation of mycorrhiza, *Eriophorum* spp., *Narthecium ossifragum*, *Myrica gale*, *Pinguicula vulgaris*, *Hydrocotyle vulgaris*, and *Drosera* spp. were cited as entirely free from fungus infection. In the light of modern researches this list of "fungus-free" species is a useful illustration of the need for an improved technique in investigations of this kind. The author confirmed Frank's records as to the absence of root hairs in Ericaceae, observing as a noteworthy fact that members of this group formed mycorrhiza even when growing in sandy soil deficient in humus.

Höveler sought experimental proof that many plants living on humus soils could use the organic compounds present, but his researches were inconclusive in respect to this and were not confirmed by contemporary workers.

Later, Reinitzer (1900) and Nikitinsky (1902) carried out investigations on the decomposition and utilisation of humus, but their results threw little light upon the nutritive relations in mycorrhiza plants.

The attention attracted to *Neottia* and other non-chlorophyllous species in respect to their mode of nutrition and the possession of mycorrhiza has already been noted.

Further contributions to the literature dealing with these so-called saprophytes were made by Johow in 1885 and 1889, and by Groom in 1894 and 1895.

Johow's admirable papers on "holosaprophytes," although they dealt but briefly with the mycorrhiza of the species described, had an important indirect bearing on the subject by reason of the special significance of the nutritive relations in the case of vascular plants lacking chlorophyll. The first paper dealt with the anatomy and embryology of a number of holosaprophytic species collected by the author in Venezuela and the Lesser Antilles; the second attempted a comprehensive account of the whole group from a biological standpoint and included descriptions of many species previously imperfectly known, together with a review of all the knowledge then available concerning these curious plants (Plate I, Figs. 5–10).

The use of the term holosaprophyte for non-green vascular species other than obvious parasites is sanctioned by custom and need not now be discussed. As used by Johow, the term was applied to non-green vascular plants growing in a substratum rich in humus.

At the time when these observations were made, practically nothing was known concerning nutrition in these plants, although certain conclusions were inferred from their usual habitat in humus-rich soils and from the fact that those investigated showed regular and characteristic fungus infections of the subterranean organs. Information as to seed germination was fragmentary and unsatisfactory, nothing was known as to their behaviour when grown under strictly controlled conditions, as to the time and manner of fungus infection or the possibility of raising seedlings free from mycorrhiza. It may be added that the field of experimental enquiry in respect to these matters still remains almost unexplored.

Only in the case of one genus mentioned by Johow, viz. *Gastrodia*, has subsequent work thrown light upon the actual method of nutrition and illuminated the nature of the "symbiotic" relation between the vascular and non-vascular constituents.

Non-chlorophyllous species other than parasites were recorded by Johow in 43 genera of Phanerogams—29 monocotyledons, distributed in the three families, Orchidaceae, Burmanniaceae and Triuridaceae, and 14 dicotyledons, all members of Ericales or Gentianaceae. In view of the disparity of numbers between the two groups, it is perhaps significant that the parasitic habit is not known to occur among monocotyledons.

Geographically, the regions richest in species were found to be Malay and equatorial South America. In general, plants were found only in moist shady woods, preferring soils rich in vegetable remains. The majority grew in a loose, spongy substratum of decaying leaves, but certain Orchids—*Epipogum aphyllum, Corallorhiza innata* and *Neottia nidus avis*—were also recorded by Johow from clayey soils watered by drainage from humus-rich deposits, and *Hypopitys* was reported exceptionally from grassy places in sand dunes, presumably poor in humus (cf. *Monotropa hypopitys* near Southport, p. 5).

Some of the tropical species grew upon trunks of fallen trees and in similar situations, and a few were epiphytes on living trees. The origin of the humus did not ordinarily appear to be important, but *Hypopitys hypophagea* was specially associated with organic detritus derived from Fir and Beech, and *Dictyostegia orobanchoides* was recorded on a substratum consisting of the dead roots of Palms.

All the species were found to resemble *Neottia* and *Monotropa* in the possession of well-developed subterranean absorbing systems with a scanty development of aerial shoots limited to the flowering axes. The morphology of the absorbing organs showed much variety in detail, e.g. single tubers or rhizomes with or without roots, rhizomes or roots with "coralloid" branching. Three species of *Galeola*, viz. *G. cassythoides, G. foliata* and *G. altissima* are remarkable exceptions, being branched climbing plants which reach a length of 50 to 120 feet.

Anatomical peculiarities were already known in certain genera. The existence of these was confirmed and found to be characteristic of all members of the group. Absence of chlorophyll, replacement of chloroplasts by leucoplasts or chromoplasts, the absence of a main root and the frequent development of the "coralloid" or "bird's nest" type of subterranean organs, the absence or feeble development of root hairs, and finally, the almost invariable development of mycorrhiza, to which Johow, in his first paper had thus alluded:—"Das constante Auftreten eines Parasiten in den Wurzeln dreier Pflanzenarten," these being *Neottia, Corallorhiza* and *Monotropa*.

Minor abnormalities of structure appeared in the roots of most species. Scale leaves of various types were borne upon the rhizomes and tubers, and also upon the aerial shoots. The latter were white or yellowish white in colour or showed a marked development of coloured pigment.

Johow recorded the complete absence of stomata as an outstanding peculiarity of shoot structure in these plants, correlation of which with absence of an assimilating and transpiring mechanism being well shown in Burmanniaceae, a group including species with foliage leaves of normal structure. All the species described produced minute seeds with rudimentary embryos.

With one exception, all the saprophytic species known to Johow formed typical mycorrhiza in the roots and (or) rhizomes, endotrophic in the great majority, but ectotrophic in *Monotropa* and its allies. The only exception belonged to the genus *Wullschlaegelia*, an Orchid closely related to *Neottia*, recorded by Johow as showing no trace of regular fungal infection. In view of the occasional presence of single hyphae mentioned by this author, a re-investigation of the species is greatly to be desired.

The distribution of mycelium in the cortical cells of the roots and rhizomes was described as varying with the species; the cytology of the infected tissues was not investigated by Johow.

In discussing the rôle of the mycorrhizal fungus in nutrition, reference was made by the author to the interesting observations of Schimper on the aerial roots of epiphytic Orchids. As is well known, certain species, e.g. *Isochilus linearis*, have aerial roots of more than one kind. Observations on those roots which ramified upon the surface of the bark and organic debris showed that fungus infection was confined to the cortical tissues upon the lower side, i.e. to that in contact with the substratum, while the aerial roots which hung freely in the air remained free from fungus infection. This distribution of infection was regarded as evidence that the mycelium was related in some way with the intake of food material—"Durch diesen Befund wird wohl deutlich angezeigt, dass die Mykorhiza in Beziehung zur Nahrungsaufnahme steht."

Johow accepted the existence of hemisaprophytic species as an established fact, instancing the works of Solms Laubach (1867–1868), Kerner (1887) and Koch (1887) on the subject. As in the case of holosaprophytes, the claim that certain species are hemisaprophytic in nutrition was based on the character of their habitats and upon certain structural features resembling those found in their nonchlorophyllous allies. No experimental evidence was produced that any of these plants could directly utilise the organic residues in humus soils.

In a paper dealing with the same subject Groom (1894) defined saprophytes as "plants which are dependent for their existence on the presence in the substratum of decaying organic matter. Like parasites, they may be divided into those which possess chlorophyll (hemisaprophytes) and those which have none (holosaprophytes)." The scanty experimental evidence for the existence of hemisaprophytes was noted by Groom, however, who pointed out that current views on the subject were largely speculative.

After reviewing the work of earlier observers, including a detailed summary of the conclusions reached by Johow on the subject, the author gave a detailed account of the morphology and histology of four Orchidaceous holosaprophytes from the Indo-Malay region, viz. *Galeola gavanica*, *Aphyllorchis pallida*, *Lecanorchis Malaccensis* and *Epipogum nutans*. Of these previously undescribed species, the three first-named possessed rhizomes with unbranched or feebly branched roots and scale leaves; the last-named species was rootless. In all four species structural features directly related to absorption and to the elimination of excess water from the subterranean organs were described by Groom. In all there was found extensive fungal infection of the roots, and (or) rhizomes, and (or) scales, with a development of endotrophic mycorrhiza similar to that in *Neottia*.

Two other Orchid species, *Corysanthes* sp. and *Spiranthes australis*, were also examined for evidence of hemisaprophytism. Both were terrestrial plants with foliage leaves and formed endotrophic mycorrhiza. The habit, histology and affinities of the former and the affinities and habitat of the latter were believed to favour the view that both species were hemisaprophytic, i.e. that they utilised the soil humus directly as a source of organic food material. The conclusions reached respecting the relation of fungus infection to metabolism were as follows:—"Essentially connected with the absorbing organs are the mycorrhizal hyphae, which are present in all known saprophytes (except *Wullschlaegelia*, according to Johow). The function of mycorrhiza is still a mystery. All that is known is that certain plants have on or in their roots mycorrhizal hyphae, that mycorrhiza is absent from roots grown in soils devoid of humus, and that in the case of endotrophic mycorrhiza the fungus does not kill the protoplasm of the cells in which it dwells." It may be noted that this observation was made nearly ten years subsequent to the publication of Frank's theory of beneficial symbiosis. Indeed, in this paper the experimental observations of the latter are dismissed as inadequate and uncritical, although the author, reviewing the distribution of the mycorrhizal fungus in the absorbing organs and its relation with the root cells of these non-chlorophyllous Orchids, believed that his own observations supplied confirmatory evidence for the view "that mycorrhizal hyphae exert a beneficial effect on the host, and that the fungus is not merely a passive companion."

Protolirion paradoxum, a representative of a new genus of monocotyledons described by Groom in 1895, was believed to show affinities with both Liliaceae and Triuridaceae, the root resembling that of members of the latter group in the possession of a well-developed cortex with endotrophic mycorrhiza. The evidence derived from other structural characters was regarded also as indicative of a saprophytic habit.

In 1895 Groom contributed a paper on *Thismia aseroë*. This plant is a member of a genus of curious "holosaprophytic" species belonging to the family Burmanniaceae. It is a native of Malaya and was discovered and first described by Beccari. The plant body consists of a branching system of cylindrical, leafless axes which ramify horizontally in a substratum of decaying leaves. At the flowering period, erect scale-bearing branches bearing terminal flowers arise from the terrestrial axis. The anatomical structure of the creeping axis is anomalous and its exact morphological nature is not evident.

The external and cortical regions of the axis were found to be extensively infected with mycelium, the mycorrhizal tissue showing a high degree of differentiation with regard to the condition of the intracellular mycelium. In the more superficial layers were found slender hyphae with densely staining protoplasm, in the deeper layers (medio-cortex) the cells contained "*conspicuous, dead, yellow, mycelial masses, consisting of portions of distinct hyphae which are connected by slender portions of defunct hyphae with one another*." Between these two regions was a single layer of cells containing "very slender hyphae, often spirally twisted, which suddenly swell out into intercalary bladder-like bodies often filled with densely stainable protoplasm."

In certain tissues the cytology of the infected cells was peculiar and showed a remarkable development of intercalary hyphal swellings in which degenerative changes ultimately took place. In such cells starch disappeared after penetration by the fungus and did not reappear.

The physiological significance of these facts is very fully discussed by the author, who concluded from comparative observations that there was an "interchange of material between Fungus and host, and that the material is manufactured by the two symbionts respectively."

Frank's designation of cells showing degenerating mycelium in the endotrophic mycorrhiza of Orchids as "Pilzfallen" and of the plants possessing them as "pilzverdauende Pflanzen" was severely handled by Groom in this paper on the grounds that the fungus demonstrably abstracted carbonaceous food from the plant cells in the early stage of infection, and that no *direct* evidence existed that it provided compensation in the form of protein material for the vascular host. "Although the hyphae of endotrophic mycorrhiza in the medio-cortex die soon, the root (or rhizome) cannot be said to act like the digestive organs of an insectivorous plant, because the protoplasm of the hyphae is manufactured partly at the expense and through the agency of the host." Moreover, the invading organism is not killed, indeed there was some evidence that in *Thismia* the hyphae in the outer cortical tissues "acted as haustoria for those outside."

Frank's views as to the different physiological significance of endotrophic as compared with ectotrophic mycorrhiza were also criticised by Groom, who concluded that "the distribution of the two forms of mycorrhiza and the occurrence of transition stages between their extreme forms, militates against the view that the physiological significance is not the same in both."

Of the alternative views that mycorrhiza is "a highly adapted and symbiotic community beneficial to both symbionts" or a "pure matter of infection of a plant by a Fungus," with a constant struggle between host and tentative parasite—Groom held that the weight of evidence was on the side of the former. His own observations on *Thismia* pointed to the absorption of carbohydrates by the fungus with some compensating return in the form of organic compounds of nitrogen from the humus made available to the host-plant. His observations pointed to a similarity of functions in ectotrophic and endotrophic mycorrhizas and provided no support for Frank's hypothesis in respect to this.

The account of *Sarcodes sanguinea* published by Oliver (1890) served to emphasise once more the remarkable biological features shown by members of the Monotropeae. This curious non-chlorophyllous species had been collected in Pine woods in two localities in the mountains of California. As described by Oliver the vegetative organs of the plant are represented only by a mass of coralloid roots, from which at flowering arises a massive inflorescence axis about 35 centimetres high, closely invested by fleshy scales. As in other members of the Monotropeae chlorophyll is not formed, but the aerial parts of the plant are coloured brilliantly owing to the presence of a soluble red pigment. The root system consists of a brittle mass of densely branched fleshy roots, each lateral member arising exogenously upon the parent root. All the roots develop typical ectotrophic mycorrhiza resembling, except in minor details, that described for *Monotropa*. The apex of each root is enveloped by a sheath of closely interwoven hyphae that extends backwards over the whole surface of the root. Hyphae from the inner part of this fungal sheath penetrate between the epidermal cells of the root but do not enter them.

Members of the genus *Corallorhiza* had long attracted the attention of botanists by reason of their peculiar coralloid rhizomes, absence of roots, and complete or almost complete lack of chlorophyll. The genus contains twelve species widely distributed throughout Europe, Asia, the United States and Mexico. All are brownish or yellowish herbs of similar habit to *Neottia* and *Monotropa*, entirely lacking chlorophyll or developing only traces of that pigment in the later stages of growth. The observations made by earlier workers have been mentioned.

A short paper by Thomas (1893) dealing more especially with *C. multiflora*, a species in which the aerial parts may reach a height of eighteen inches, was followed by one from Jennings and Hanna (1898) on *Corallorhiza innata*. Both papers described the curious trichome-bearing papillae upon the surface of the rhizome of this Orchid, Thomas regarding the hairs as organs of attachment that showed little evidence of any parasitic function, Jennings and Hanna interpreting them as "fungus traps" facilitating infection of the rhizome by the mycorrhizal fungus.

The cytology of the mycorrhiza cells is inadequately described in both papers, neither of which made a large contribution to the knowledge of endotrophic mycorrhiza in non-green plants. In respect to the biological habit, Thomas rejected the hypothesis of parasitism—"instead of being a root parasite as has been supposed, the plant depends chiefly on the symbiotic condition for its food and this is taken by the hyphae from the decaying vegetable matter about." Jennings and Hanna accepted the view put forward by

Frank that "the fungus is a living organism captured for the benefit of the host plant," and considered that the latter was "at least by far the larger shareholder in the symbiotic relationship, if it can be regarded as such. More probable seems the view that there is no symbiosis, but that the fungus is captured and utilised by the Orchid without any compensating benefit." No direct evidence in support of either of these views was offered and no experimental work was attempted on either species. In the case of *C. innata* from the Eastern Alps, Jennings found evidence that the mycorrhizal fungus was a Hymenomycete, possibly *Clitocybe unfundibuliformis*, sporophores of which were constantly associated with the plants, but fruit bodies of *Tricholoma* sp., *Cortinarius* sp. and of a subterranean Hymenomycete were also found in close proximity.

Before leaving the subject of *Corallorhiza*, mention may be made of the observations of Lundström (1889) on *Calypso borealis* in Sweden. This rare Orchid possesses a "coralloid" rhizome with a local development of endotrophic mycorrhiza resembling that of *Corallorhiza*. The observation recorded by Lundström that mycorrhiza is not always formed in *Calypso* was confirmed later by MacDougal (1899 *a*, *b*), who observed great variability in plants of this species, not only in respect to the formation of mycorrhiza and "coralloid" branching of the rhizome, but also in leaf and flower characters. It was noted also by Lundström that ripe fruits were difficult to find and that seedlings were rarely seen in Nature. Moreover, attempts to raise plants from seed were not successful.

The possession of endotrophic mycorrhiza by numerous species, other than those belonging to the four families, Orchidaceae, Ericaceae, Epacridaceae, and Empetraceae, originally cited by Frank, had been established by the researches of Schlicht and Johow. Janse (1896–7) extended the already lengthy list by recording the wide distribution of this type of root infection among tropical plants. His attention was attracted to the subject when working on the fungus parasites of Coffee in Java, and his observations were made on plants from the forest of Tjibodas and the Botanic Garden and its neighbourhood at Buitenzorg. His researches extended over a wide field, including Bryophytes, vascular Cryptogams, Gymnosperms, Monocotyledons, and a large number of woody Dicotyledons. Of the 75 species studied, 69 showed typical endophytic infection without damage to the cells of the host, affecting the roots or rhizomes according to the habit of the species.

Certain special organs mentioned by earlier workers were studied in greater detail by Janse. His observations on this matter have a bearing on those of subsequent workers and may be briefly summarised as follows.

Soon after infection the endophytic mycelium formed terminal swellings—"vésicules"—varying in shape and size from spherical bodies 20μ in diameter to elongated sacs 100μ × 27μ or larger. They contained much granular cytoplasm, and when mature, became gorged with reserve products, especially oil, and often acquired thicker walls of a brown colour. They were formed both within cells and in intracellular spaces and were regarded as analogous with similar structures recorded by Bruchmann (1874) and Goebel (1887) for *Lycopodium* sp., Kühn (1889) in *Angiopteris*, Schlicht (1889) in *Paris* and *Ranunculus*, Groom (1895) in *Thismia aseroë* and Poulsen (1890) in *Sciaphila* sp. In common with a majority of other observers Janse was inclined to regard them as of the nature of resting spores which functioned as a means of asexual propagation for the endophyte. In general, "vésicules" were absent from Orchid mycorrhiza, but in the case of two genera, *Platanthera* and *Epipactis*, Mollberg (1884) had recorded structures believed by Janse to be analogous to them.

In describing organs of another type formed by the endophytes, Janse was on less sure ground, and it is clear from his account and from the figures illustrating it that he found it difficult to account satisfactorily either for the structure or functions of these so-called "sporangioles." They were ubiquitous in distribution but were absent from the mycorrhiza of Orchids, where they appeared to be replaced by other structures, and also from that of seven tropical species other than Orchids. They were always intracellular, in distribution; they were formed in one or more layers of cells in the more deeply situated cortical tissues; they were rounded or irregular in shape and showed a mammillated structure interpreted as due to the inclusion of "sphérules" filled with granular material (Pl. II, fig. 16). Janse was evidently disposed to regard the "sporangioles" of his tropical species as analogous to the "corps jaunes" of Wahrlich in Orchids, the "dichte Massen" of Schlicht in other species, and also to the structures "which break up into bacteriods," described by Groom in *Thismia*. In view of the elucidation of their real nature by subsequent workers, it is unnecessary at this point to discuss them further.

Although Janse's paper was, in the main, an account of the morphology and anatomy of mycorrhiza in tropical plants, he did not ignore its physiological significance, and frankly admitted the necessity for isolation and study of the endophytic fungi. He referred to the unsatisfactory character of the attempts made by previous workers, and to his own unsuccessful efforts in the case of the endophyte of Coffee.

His speculations on the biology of the relationship in mycorrhiza were obviously affected by the fascinating and successful researches of Winogradski (1895) on *Clostridium pasteurianum* and by the work of Beijerinck (1888, 1890) and Laurent (1891) on the nodules of leguminous plants. Pointing out that the assistance rendered to the vascular plant by the fungus might take the form of inorganic salts, of organic substances other than nitrogenous, or of nitrogenous material, he marshalled the indirect evidence for the fixation of atmospheric nitrogen by the endophytes with the subsequent transfer of nitrogenous material to the host plant, and concluded a somewhat unconvincing argument for a reaction towards oxygen, resembling that in *Clostridium*, as follows:

> L'endophyte étudié est un champignon aérobie facultatif de même que le *Rhizobium* et le *Frankia*. Il habite la grande majorité des plantes les plus diverses, et se loge dans les couches internes de la racine où il vit aux dépens des hydrates de carbone de son hôte. En pénétrant dans les tissus vivants, il cherche surtout à éviter l'oxygène. Dans ces conditions il a la faculté de fixer l'azote atmosphérique. La plante hospitalière s'empare de la plus grande parte des matières azotée que prépare le champignon et se fait payer ainsi la nourriture hydrocarbonée et la protection qu'elle lui accorde.

It may be mentioned that the experimental results furnished by cultures of Coffee in sterilised soils in no way supported these opinions. On the contrary, the experimental plants showed equally vigorous growth in treated and untreated soils.

The interest excited by the character of the nutrition in non-chlorophyllous species and their green allies suspected of a saprophytic habit was evidenced in a series of papers by MacDougal (1898–1899). The author aimed at extending the knowledge of mycorrhiza—its occurrence, the physiological relation between the symbionts, and the effect of mycorrhizal adaptations upon development with reference to survival value in the species. Two aspects of the subject engaged his attention—the recording of fresh observations upon a number of species, and the expression of his own views respecting the saprophytic habit in general. Among the plants examined were *Pterospora andromedea*, a holosaprophytic member of the Monotropeae, eight Orchidaceous species recorded as *hemi-saprophytes*, including *Calypso borealis*, to which reference has already been made (see p. 42), together with *Podophyllum peltatum*, *Sarracenia purpurea* and other autotrophic species found to be free from mycorrhiza. Of new observations may be noted: the presence of stomata in species of *Epipogum*, *Aphyllorchis*, *Lecanorchis*, *Cotylanthera* and *Pterospora*, from which as from other non-green saprophytes their absence had been recorded by earlier observers, and the penetration of the endophyte into tissues other than those of absorbing organs. It does not appear that the cytology was studied in detail, but in *Peramium* (*Goodyera*) *repens*, intracellular mycelial masses were described as structures of an "absorptive character," not in any sense homologous with or analogous to the "sporangioles" of non-orchidaceous species as indicated by Janse. Despite which fact, the author appears to have regarded both "sporangioles and hyphal clumps" in general as "organs of nutrition." Although not figured, there can be little doubt of the real nature of these intracellular structures described by MacDougal; in common with "sporangioles," they will be considered more fully in a later paragraph.

The opinions expressed by MacDougal respecting nutrition of saprophytic species were founded on anatomical considerations and must be regarded as entirely theoretical. Emphasis was laid on the fact that the young seedlings of ordinary autotrophic plants are saprophytic in nutrition. The existence of this habit in the adult can be explained therefore by assuming retention of a juvenile character together with loss of chlorophyll and the development of certain structural modifications of the absorbing and transpiring organs.

MacDougal applied the term *holosaprophyte* to all species lacking chlorophyll, and *hemisaprophyte* to those showing only slight modification in structure. Accepting this customary definition, he pointed out that only three methods existed by which the requisite organic materials could be obtained by such plants:

(*a*) By adoption of the carnivorous habit. Since no carnivorous plant was known in which this habit was obligate, it followed that no species has attained complete saprophytism in this way.

(*b*) Directly, by modification of the absorbing organs, to permit the intake of organic compounds. Complete saprophytism of this kind was known only in fungi, bacteria, and—assuming the correctness of Johow's observations—in one vascular plant, the Orchid, *Wullschlaegelia aphylla*.

(*c*) Indirectly, through the intervention of fungi which had invaded the absorbing organs. In this case it was assumed that "the walls of the fungus have developed a capacity for the osmotic passage of organic material.... With regard to the higher plant, therefore, I have temporarily termed this adaptation *symbiotic saprophytism*," such condition being regarded as "the natural result of the supplemental capacities of two organisms brought into nutritive contact chemotropically."

MacDougal found no support for the theory advanced by Janse that root fungi were negatively chemotropic to oxygen, and under anaerobic conditions bore a relation to the host plant similar to that shown by the nodule bacteria in legumes. Moreover, he distinguished two types of endotrophic mycorrhiza, "one adapted for nitrogen fixation, and a second for the absorption and modification—perhaps oxidation—of the soil products before liberation in the tissues of the higher plant." Unfortunately, in spite of the interest of his observations, MacDougal was not able to support his views on nutrition by reference to experiment, or to contribute any new facts respecting the identity of the fungi concerned in the formation of mycorrhiza.

Evidence of a growing interest in cytological problems was provided by a new contribution to the already extensive literature on *Neottia*. In the preface of this paper Magnus (1900) referred to the dearth of observations on plant material relating to cell pathology, the contributions of Vuillemin (1890) on mycorrhiza, and of Tubeuf (1895) on diseases of plants containing little information as to changes in nuclear structure or in the finer cytological details.

It was with the view of filling this gap that Magnus undertook a comparative study of the cytology of the infected cells of Orchid roots with special reference to *Neottia nidus avis*. His observations correlated and explained those of earlier observers in respect to certain features; the paper recording them forms a starting point for modern work on the subject, and the facts described in it have since become a commonplace of the literature dealing with Orchid mycorrhiza.

A correct interpretation of the slimy brownish masses present in many of the root cells of *Neottia* and other Orchids, due in the first instance to Cavara (1896), had been confirmed by Chodat and Lendner (1896), who, in an account of the mycorrhiza of *Listera cordata*, had described the degenerative changes of mycelium and other cell contents which took place in the older tissues. Dangeard and Armand (1898) had made a similar contribution in respect to *Ophrys aranifera*. These views had replaced the incorrect and often confused interpretations offered by Drude, Reinke, Mollberg and Wahrlich.

In *Neottia,* Magnus recorded invariable infection of the vegetative organs involving three or four layers of cortical tissue in the roots, and sometimes as many as six layers of sub-epidermal cells in the stems. Two types of infected cell were described, differing markedly in structure and physiological reactions and unconnected by transition forms—"Pilzwirthzellen" in which the mycelium retained its identity and persisted in the active condition, and "Verdauungzellen" in which it underwent rapid degeneration.

In both types of cell the hyphae on entering became clearly associated with the cell nucleus and branched extensively. In both, subsequent to infection, the mycelium was completely invested by the cytoplasm of the host cell, the latter showing a similar degree of vitality in infected and uninfected cells. In the "Pilzwirthzellen," the cell cavity became filled with mycelium showing a differentiation to thick-walled hyphae of large diameter at the periphery surrounding a central region filled with fine hyphae, the whole enveloped by the cytoplasm of the host cell. The "Verdauungzellen" in the earlier stages were filled with closely wound skeins of thin-walled protein-rich hyphae—the "Eiweisshyphen." Later, the nuclei of these cells underwent remarkable changes of size, shape and chromatin content, indicating marked functional activity, and the mycelial contents, surrounded by and including a proportion of the cell cytoplasm, became "clumped" around them, the whole contents of each cell eventually forming an opaque structureless mass—the "gelblichen Stoffes" of Schleiden and "corps jaunes" of Wahrlich and subsequent observers.

"Clump" formation involved the whole or a part of the cell contents and was followed by complete degeneration and digestion of the mycelium. As described by Magnus, each "clump" became surrounded by a layer of cellulose-like material secreted by the cytoplasm. Whether this actually was the case or whether the substance of the clumps—apart from the cytoplasm—consisted entirely of alteration products of the digestion process, remained somewhat uncertain.

Certain features of interest emerged from this first clear and accurate account of intracellular digestion in Orchids. The two types of cell in *Neottia* were described as showing a regular arrangement, the "Pilzwirthzellen" occupying the middle part, and the "Verdauungzellen" the outer and inner layers of the infected region.

Magnus reported a similar differentiation of the mycorrhiza cells in other Orchids, varying in degree in different species, and proposed a classification of Orchid mycorrhiza on this basis.

No transitions were observed between the two types of cell. In the one case, the hyphae retained their activity throughout the life of the root, in the other they invariably suffered complete degeneration. In the former, the fungus grew as a parasite, caused injury to the cell, and eventually formed resistant hyphae especially adapted to survive the winter after the death of the roots; in the latter, the products of digestion were at the disposal of the host plant. Physiologically the two types of cells were regarded as representing profit, in the one case exclusively to the fungus, in the other exclusively to the vascular plant.

Much attention was devoted by the author to the observation and interpretation of the nuclear changes observable in infected cells—their increase in chromatin content and stainability, and, in the case of "Verdauungzellen," the amoeboid changes of shape. He does not appear to have related these appearances directly to the digestive activities of the cells and was obviously puzzled by the close association of the "parasitic" hyphae and their "haustorial" branches with the cell nucleus, in relation to current views on the significance of the latter as the centre of nutritive activity in the cell. "Dass sich parasitäre Pilze, mit ihren Haustorien oft an den Zellkern legen und in seiner Nähe eigenthumlich verzweigen, gestattet Keinen Rückschluss auf die Bedeutung des Kernes als Nahrungscentrum der Zelle."

His observations confirmed the prevailing opinion that the Orchid endophytes, with their scantily developed soil connections, did not play an important part in the absorption of nutritive material from the soil, finding the food supplies necessary for growth within the host cells.

Magnus compared the biological relation in the mycorrhiza of Orchids with that in leguminous nodules, regarding the "Verdauungzellen" as a mechanism whereby the higher plant benefited and the "Pilzwirthzellen" as one whereby certain vegetative parts of the endophytes, having escaped digestion, were returned to the soil as resting structures after the death of the roots.

Chodat and Lendner (1896) had rejected the view of a symbiotic relation beneficial to the Orchid plant in the case of *Listera cordata*, and regarded the endophyte as a relatively harmless parasite.

Regarding the condition in each mycorrhiza cell of *Neottia* as a life and death struggle for the individual symbionts, Magnus envisaged the possible existence of other and extreme types of symbiosis in other Orchids—destruction of the invaded cells at one end of the series, complete digestion of the invading mycelium at the other. A discussion on the biological significance of mycorrhiza in non-chlorophyllous plants concluded with a warning against accepting premature conclusions regarding the biological relations in view of the ignorance that existed at the time in respect to the general relations of these plants with their environment.

He also alluded briefly to the mycorrhiza of *Andromeda polifolia*, which had been figured by Frank without reference to the condition of the intracellular mycelium, and to his own observations on *Erica* and *Vaccinium* which suggested a differentiation into "Pilzwirthzellen" and "Verdauungzellen" in these genera of Ericaceae. It was also pointed out that a figure of the root cells in *Calluna* published by Pfeffer (1897) showed cells of the former type only.

Magnus contributed nothing fresh to knowledge of the identity of the root fungi of *Neottia* or other Orchids.

Lendner (1895) had re-examined the endophytes of *Platanthera* and *Vanda* from this point of view, and had accepted Wahrlich's conclusions with regard to their systematic position. Chodat and Lendner (1896) also reported that the endophytic fungus of *Listera cordata* resembled the *Nectria* isolated by Wahrlich from species of *Vanda*.

The observations of Shibata (1902) on certain Japanese plants form an interesting corollary to the work of Magnus on Orchids. Cytological investigation of two species of *Podocarpus* and of *Psilotum triquetrum* showed that the richly developed intracellular mycelium in the cells of the roots and rhizomes respectively underwent rapid digestion, with subsequent disappearance of the products. At the onset of digestion, the nuclei of infected cells exhibited features indicating great metabolic activity—increase of size and chromatin content, amoeboid changes of shape and multiplication by amitotic division, the latter being regarded by the author, not as a symptom of degeneration, but as a means of rapid increase of the centres of functional activity in each cell. After digestion, the nuclei and host cells reverted to their ordinary condition.

The mycorrhiza of *Podocarpus* was developed as serial rows of root tubercles formed by increase in number and modification of the lateral roots; subsequent to digestion the tubercles disintegrated and were not liable to reinfection. According to Shibata, the material remaining in the cells after digestion and absorption consisted of the remains of the hyphal membranes, surrounded by an amyloid-like material. The "vésicules" formed by the endophyte were regarded as comparable with the swellings on vegetative hyphae that appear under certain cultural conditions.

The presence of active proteolytic enzymes in the tubercles of *Podocarpus* was determined by extraction of the tissues in glycerine and estimation of the changes induced in fibrin by the extract. Shibata considered that the endophyte absorbed some part of the necessary carbohydrate food material from the host plant, the symbiotic relation consisting in the acquisition of certain protein substances by the latter. In view of the scanty connections with the mycelium outside the roots, he regarded it as doubtful that the plant could benefit by reason of absorption of mineral salts by the fungus, and he preserved an open mind with regard to the general method of nutrition of the latter.

In *Psilotum* the fact of infection had been recorded previously by Solms Laubach (1884), Janse (1896–7), and Bernatsky (1899). In this plant it was the rhizome which suffered invasion, and the infected cells showed differentiation to "Pilzwirthzellen" and "Verdauungzellen" as in Orchids.

Shibata's observations on *Psilotum*, as well as those on the root tubercles of *Alnus* and *Myrica*, are dealt with in later sections of the present work.

Petri (1903) subsequently repeated Shibata's observations on species of *Podocarpus* with special reference to the rôle of the organs described as "sporangioles." Like the latter observer, he was able to prepare from the roots a glycerine extract with marked proteolytic properties.

Explanation of Plate I

Fig. 5. *a*, *Dictyostegia orobanchioides*; *b*, ditto, leaf-scale from rhizome; *c*, *Pogoniopsis* sp.; *d*, *e*, *f*, *g*, *h*, *Sciaphila schwackeana*.

Fig. 6. *a*, *Voyria tenella*; *b*, ditto, seedling from dead wood; *c*, *Voyria uniflora*; *d*, *Voyria trinitatis*.

Fig. 7. *Pogoniopsis* sp.; tr. section of lateral root.

Fig. 8. *Apteria setacea*.

Fig. 9. *Burmannia capitata*; plant growing in dead wood.

Fig. 10. *a*, *Voyria obconica*; *b*, *Burmannia capitata*; *c*, *Gymnosiegia refracta*; *d*, *Dictyostegia orobanchioides*. Original figures of plants all natural size. Reduced to about $\frac{1}{3}$.

(From Johow, *Prings. Jahrb.* vols. 16 and 20.)

PLATE I

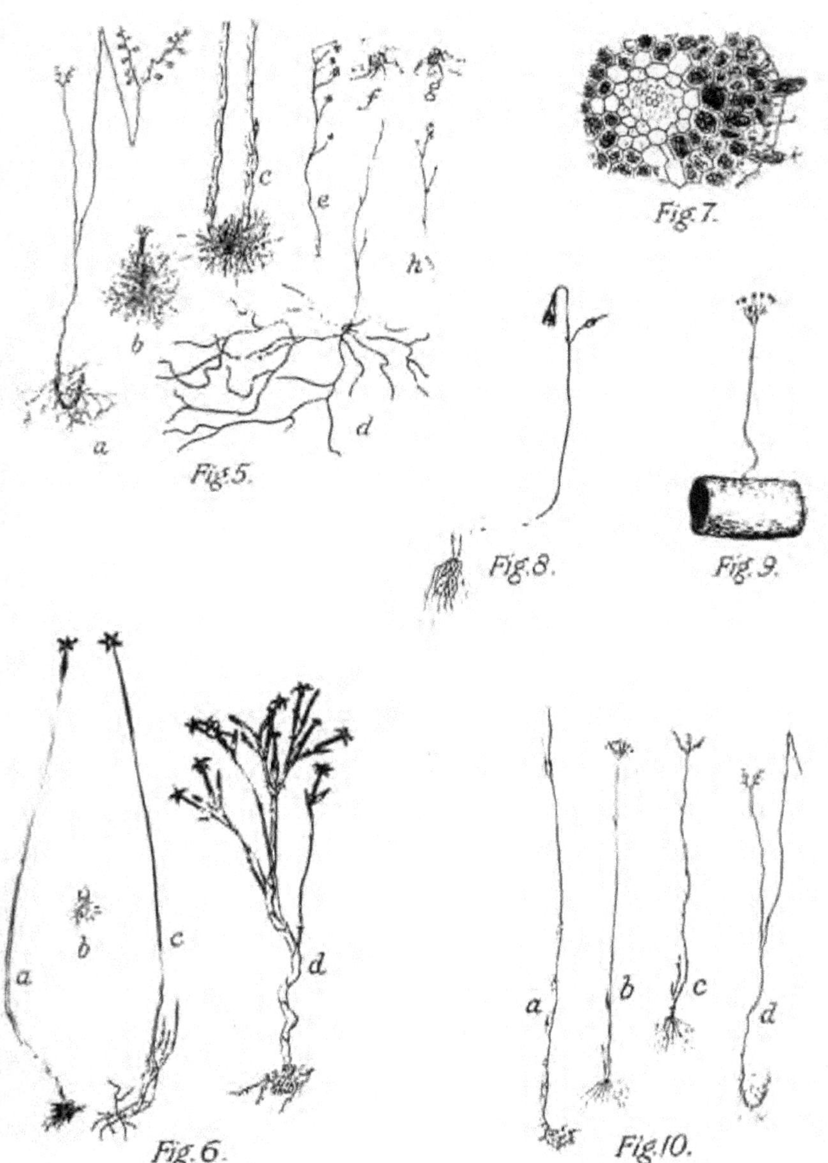

Fig. 5. Fig. 6. Fig. 7. Fig. 8. Fig. 9. Fig. 10.

CHAPTER IV

THE MODERN PERIOD

The Modern Period: 1900–1925—Sarauw—Stahl: autotrophic and mycotrophic plants; theory of nutrition—Marcuse—Mycorrhiza of Arctic and Alpine plants: Hesselman; Schröter; Wulff—Gallaud.

The Modern Period: 1900–1925.

On the whole, the views current at the beginning of the present century respecting ectotrophic mycorrhiza tended to attach rather less importance to Frank's opinions without abandoning the theory of a symbiotic relation. Thus, Tubeuf (1903) from 1896 onwards had observed the presence of functional root hairs in the ectotrophic mycorrhiza of trees and noted that a proportion of the roots were free from fungal infection. He had recorded also the not infrequent appearance of the endotrophic type of mycorrhiza in forest trees and was of opinion that the latter obtained the necessary salts independently, the root fungi functioning only as an indirect means of drawing upon the organic compounds of nitrogen locked up in humus. Observations and experiments on forest trees led Sarauw (1903) to conclude that the fungi were relatively harmless to the trees, although he regarded it as improbable that the latter derived any benefit from the association—"Dass die Pilze unseren Waldbaumwurzeln und den Bäumen Vorteil bringen sollten, ist bisher meines Erachtens weder durch Beobachtungen in der Natur, noch durch Versuche nachgewiesen worden."

Möller (1903, 1906) had reached similar conclusions. His observation that Pines made good growth on sandy soils deficient in humus had led him to dispute Frank's view that *Pinus sylvestris* does not come to maturity on normal soils if mycorrhiza formation is hindered owing to lack of humus and the absence of the appropriate fungus; moreover, his experiments to test the possibility of nitrogen fixation by the root fungi of Pine and Oak had yielded negative results for both these trees.

A fresh attempt to survey the whole field and correlate the problems of nutrition presented by mycorrhiza plants with those in parasites and insectivorous plants was made by Stahl at the opening of the new century (Stahl, 1900). This paper, the most comprehensive study of mycorrhiza from the biological point of view since the

4-2

publication of Frank's theory of symbiosis in trees, has been freely quoted in the text-books and is probably one of the best-known contributions to the literature of the subject. Two aspects of Stahl's work demand attention: firstly, the new hypothesis put forward by him to explain the distribution of fungus infection in vascular plants and its beneficial effect upon the hosts, and secondly, the character of the experimental evidence offered in support of his opinions.

Stahl's theory of nutrition was supported by the following arguments:

(1) Both vascular plants and fungi make heavy demands upon certain essential mineral salts in the soil; moreover, investigation shows that fungus mycelium is a very efficient mechanism for removing soluble salts from the soil. There must be, therefore, a powerful struggle for the essential mineral salts in soils, more especially in those rich in humus in which fungi are specially abundant.

(2) The higher plants best fitted for competition under these conditions will be those with extensive root systems and abundant root hairs, in which a rapid transpiration current is facilitated by structural characters favourable for transpiration, e.g. by the presence of hydathodes, etc.

(3) Vascular plants of this type growing in humus soils are usually free from fungus infection, e.g. Elder (*Sambucus nigra*), members of the Cyperaceae, and various Ferns. Obligate mycorrhiza plants, on the other hand, have commonly a sluggish transpiration current, whether due to inefficient absorption, or to an ineffective mechanism for the elimination of water from the shoot, or to various interactions between these factors. On humus soils in particular, they cannot compete with rapidly transpiring species or with fungi without the assistance of the symbiotic mycelium in their mycorrhiza. The great benefit derived in this way becomes evident when it is recalled that the development of the root fungi reaches a maximum in the autumn, i.e. at the season when transpiration is slowest.

The major part of the paper consisted of an elaboration of the thesis just outlined, namely, that the incidence of fungus infection was directly related to the difficulty of procuring mineral salts, and hence, to the efficiency or otherwise of the mechanism for their absorption, it being assumed throughout that a large intake of water and a rapid transpiration current involved, of necessity, a correspondingly great absorption of nutritive salts from the soil. As a matter of convenience Stahl distinguished two groups of plants—those with and those without mycorrhiza, the former group including *obligate mycorrhiza plants*, which depended always upon root infection for their proper nutrition, and *facultative mycorrhiza plants*, which were infected or fungus-free according to the character of the substratum. In reference to nutrition, all mycorrhiza plants were described as *mycotrophic*.

Reviewing his own observations and also those of other workers on the distribution of fungus infection, and neglecting certain contradictory facts, Stahl concluded—"dass die Mycorhizenbildung höchst wahrscheinlich mit der erschwerten Nährsalzgewinnung in irgend einem näheren Zusammenhang steht." He stressed the intensity of the competition for essential salts in the humus soils with which many of the best known mycotrophic species were specially identified, and proceeded to test his theory of nutrition by seeking evidence that mycotrophic plants in general showed anatomical and morphological characters related to their symbiotic mode of nutrition.

The details of this investigation on mycotrophic species—the relation of transpiration to sugar or starch in the leaf cells, the distribution of hydathodes and other mechanisms for facilitating the flow of water, the observed deficiency of nitrates and other salts in the tissues—need not be considered in this review. Many observations of interest were included in the argument and may have a real significance of the kind postulated by Stahl. Nevertheless, they cannot be regarded as providing convincing evidence of the general proposition that the mycotrophic habit is a special adaptation giving to certain species compensation for the handicap imposed upon them by a sluggish transpiration current. In the light of modern work on absorption it is indeed doubtful whether the assumed direct correlation between intake of water and absorption of mineral salts exists. The evidence points to the two processes being entirely distinct, i.e. that a rapidly absorbing and rapidly transpiring plant is not necessarily specially favoured in the competition for mineral salts.

Furthermore, some of the experimental evidence offered by Stahl is open to interpretations other than those suggested. For example, a method adopted was to grow comparative cultures of various species in suitable soils and in similar soils sterilised by exposure to ether vapour for five days. The experiments were designed to illustrate the potent effect of fungus competition by removal of this factor in the soils sterilised with ether, the beneficial effects observed in seedlings being attributed to the decreased competition for salts. Whether or not the differential growth actually observed in the roots of the two sets of experimental plants was due, as believed by Stahl, to the retarding effect upon growth of a relatively higher concentration of salts due to the absence of mycelium in the sterilised soils, the bearing of this observation upon the biological function of mycorrhiza is indirect, and it in no way provided experimental proof that the presence of mycelium in the root tissues of mycotrophic plants facilitated the absorption of salts from the soil. The increased vigour observed in seedlings in the treated soils is shown in the photograph now reproduced (Fig. 11).

Fig. 11. Stahl's experiment to demonstrate the effect of fungus competition in soil. Left, two pots with *Linum usitatissimum*; the larger plant growing in sterilised, the smaller in unsterilised humus. Right, four pots with *Sinapis alba*; the two larger plants in sterilised, the two smaller in unsterilised humus. (From Stahl, *Jahrb. f. wiss. Bot.* 1900.)

In Stahl's earlier experiments the treated soils were sterilised by heating to 100° C.; in his later work, to which the series of experiments figured belong, sterilisation was effected by exposure to the vapour of chloroform or ether. In view of modern work on soil sterilisation, it is certainly unsafe to conclude, as did Stahl at the time of the experiments, that the beneficial effect observed was due solely to decreased competition for essential salts. The effects produced by *partial sterilisation* of soil, whether by heating to 100° C.

or by exposure to volatile antiseptics, are complex in origin, and lead directly and indirectly to an increase in the amount of available plant food present. The stimulus to growth commonly noted results from this increased supply and would mask any effect due to removal of fungus competition (Russell, 1921).

By independent observation, Stahl confirmed Frank's conclusion that obligate mycorrhiza plants always gave a negative reaction when tested for nitrates, even when growing in the same soil with fungus-free species which reacted strongly. This fact has no essential bearing on the hypothesis of a differential rate of absorption in mycotrophic as compared with autotrophic plants, significant though it may be in relation to the nitrogen metabolism of the former group. His comparative observations on insectivorous plants—their distribution upon soils poor in nitrates and other essential salts and their freedom from mycorrhiza—and his conclusions that the carnivorous and mycotrophic habits appeared to be mutually exclusive, are of interest in the same connection.

It was assumed by Stahl, as by earlier workers and by his contemporaries, that fungus infection took place exclusively from the soil. Modern work upon Ericaceae, with the resulting experimental demonstration of the real character of the symbiotic relationship in this group, has proved this view to be erroneous and, incidentally, has raised doubt as to the correctness of his observations on *Vaccinium myrtillis*. In the paper now under consideration it was recorded that seeds of this species sown on peat in May gave plants the roots of which showed beautiful and characteristic mycorrhiza when examined in the following October, whereas the roots of seedlings of the same age grown upon similar soil sterilised by heat were entirely free from fungus infection:

> Hierbei stellte sich heraus, dass die Exemplare, deren Wurzeln sich in dem der Siedehitze bezw. Aetherdämpfen ausgesetzten Substrat ausgebreitet hatten, völlig pilzfrei waren. Bei den in nicht sterilisirtem Substrat wurzelnden Pflänzchen waren dagegen die für die Ericaceen charakteristischen Mycorhizen in schönster Weise zur Ausbildung gelangt.

If, in *Vaccinium* as in *Calluna*, casual infection of the seedling from the soil is supplemented by regular infection from the seedcoat at germination, Stahl's conclusions obviously require revision. This matter will be discussed again when dealing with the Ericaceae.

Working in Stahl's laboratory, Marcuse (1902) re-investigated the mycorrhiza of a number of Orchids and other species, with a view to confirming the opinions expressed by the former respecting its biological significance. His observations on green and non-green Orchids led him to conclude that the condition and distribution of the endophytes stood in direct relation with the vegetative period, general environment of the plants, and age of the roots.

The structural evidence was held to support the view that out-growing hyphae functioned in a similar way to root hairs; at the same time, the author reiterated the uncertainty respecting the ability of non-green plants to draw directly upon the humus constituents of soil, i.e. he questioned whether they were, in the strict sense, saprophytic in nutrition. Among the species examined, other than Orchids, were the following—*Botrychium lunaria*, *Linum catharticum*, *Polygala amara*, *P. vulgaris* and various members of Pyrolaceae, all of which are typically mycotrophic plants. Heinricher (1900) had previously rejected the hypothesis of a parasitic habit for the genus *Polygala*, recording *P. chamaebuxus* as an obligate mycorrhiza plant and other species of the genus as facultatively mycotrophic. The figure of a root section of *P. amara* supplied by Marcuse suggests that typical intracellular digestion occurs in the latter species, although this matter is not mentioned by the author (Pl. II, Fig. 17). A few new cases were added to the rapidly growing list of plants known to form endotrophic mycorrhiza, but in the main this paper may be regarded as consisting of special pleading in support of the Stahl hypothesis; namely, that the demand for available mineral salts in certain soils exceeds the supply, the resulting intensity of competition being the primary cause of a symbiotic relation in mycorrhiza.

A somewhat new aspect of mycorrhiza was opened up by Hesselman (1900), who examined a number of Arctic species from Bear Island, Spitzbergen, and other northern stations during a Swedish polar expedition in the summer of 1898. Well-developed mycorrhiza of various types was recorded for *Salix* sp., *Dryas octopetala*, *Diapensia lapponica*, *Taraxacum phymatocarpum*, *Arnica alpina*, *Erigeron uniflorus* and *E. compositus*. In *Polygonum viviparum* it was of typical ectotrophic structure, an exceptional condition in herbaceous plants.

The formation of mycorrhiza by arctic and alpine plants is not in itself remarkable, but its existence was so bound up with the theory of nutrition from soil humus, that, at the time of Hesselman's observations, it must have occasioned some surprise to find root infection well developed in plants growing in poor and sterile soils in which a deficiency of humus constituents was determined both by the paucity of plant remains and by the unfavourable climatic conditions for bacterial action. Hesselman described mycorrhiza in *Dryas octopetala* from stations so diverse as Spitzbergen, Nova Zemlya, North Europe, Tirol, the Apennines and the Altai Mountains; the variety *integrifolia* was said to be also typically infected.

Subsequently, Schröter (1908) figured the mycorrhiza of *Pinus montana* and various herbaceous Alpine plants including the grass *Sesleria coerulea*. In the last-named species, root infection was found to be sporadic and was recorded as possibly related to the character of the habitat. In another grass, viz. *Nardus stricta,* Schnellenberg had described mycorrhiza of the endotrophic type. In this case the incidence of infection was correlated with well-marked differentiation to long and short roots, the latter characterised by arrested growth, profuse branching, absence or poverty of root hairs, and heavy intracellular infection. Stahl (1900) had previously recorded the formation of mycorrhiza by many alpine species of Gentian.

In connection with Stahl's hypothesis that the mycotrophic habit is correlated with a difficulty in obtaining mineral salts, with feeble transpiration, and with poverty or absence of starch, it was pointed out subsequently by Wulff (1902), on the basis of his own researches on Arctic plants, that a large number of species from high latitudes possess well-developed mycorrhiza. Moreover, of sixteen species collected and investigated by this author in Spitzbergen, all were characterised by the presence of abundant sugar and the absence or scanty appearance of starch in the leaves, and exhibited a very feeble transpiration current as compared with that developed by plants in warmer climates. For these Arctic species, the most favourable conditions for transpiration appeared to be provided by temperatures about 5° C. and a relative air humidity of 60–70 per cent.

The last decade of the nineteenth century had yielded a notable accumulation of observations on the distribution, cytology and biological significance of endotrophic mycorrhiza. Stahl had put forward new views respecting the beneficent rôle of the endophytes. The observations of Magnus (1900), Shibata (1902), and Petri (1903) had confirmed Frank's observations on the mycorrhiza of Orchids and shown that intracellular digestion with subsequent absorption of the products was not confined to this group of plants.

During the opening years of the new century the extensive researches of Gallaud (1905) pushed home these conclusions, while the improvement in microtechnique provided him with the means of explaining certain inconsistencies in the records of his predecessors, especially in respect to the organs described as "sporangioles."

Holding the view that many of the earlier observations on endotrophic mycorrhiza had been made on species showing a relatively advanced and complex condition, Gallaud undertook an investigation of cases likely to exhibit less specialised relations with the view of throwing fresh light on the biology of mycorrhiza. The work planned included the examination and description of the mycorrhiza of a large number of species, including some previously examined by Schlicht and others, a study of the structure and functions of special organs formed by the endophytes, together with observations on any reciprocal action observable between fungus and host, and the isolation and cultivation of the endophytic fungi. Included with his original observations on these matters are a useful historical review and an extensive bibliography of the literature.

Gallaud's observations on the production of special organs by the endophyte are of interest, and the structures described were found in practically all the plants examined by him. In the intercellular spaces of the root cortex he noted that the mycelium commonly formed cylindrical expansions the size and shape of which were determined by those of the individual air-spaces in which they were found. Sometimes these were replaced by terminal and intercalary swellings with thickened walls and densely granular contents, identical in appearance with the "vésicules" described by Janse and other workers. Occasionally the hyphae branched repeatedly in the intercellular spaces; more often this took place within the cells, giving rise to characteristic intracellular structures named by Gallaud "arbuscules." On casual examination appearing as homogeneous granular masses filling the cells, in thin sections the "arbuscules" could be resolved into more or less complicated branch-systems or tufts of hyphae, the ultimate branchlets of which, 1μ to 2μ in diameter, were difficult or impossible to distinguish without the aid of a careful technique and high magnification.

They were terminal or lateral upon the main hyphae and showed much variation in morphological detail, varying in size and complexity of branching in different plants (Figs. 12 *a*, *b*). In certain cells "arbuscules" were replaced by granular masses often nodular in outline but without definite hyphal contours. Showing marked variability in structural details and stainability, the latter were evidently analogous to the structures described by Janse as "sporangioles" and by Petri (1903) under the name of "prosporoidi." Gallaud retained the former name without attaching any significance to them as reproductive organs of the fungus.

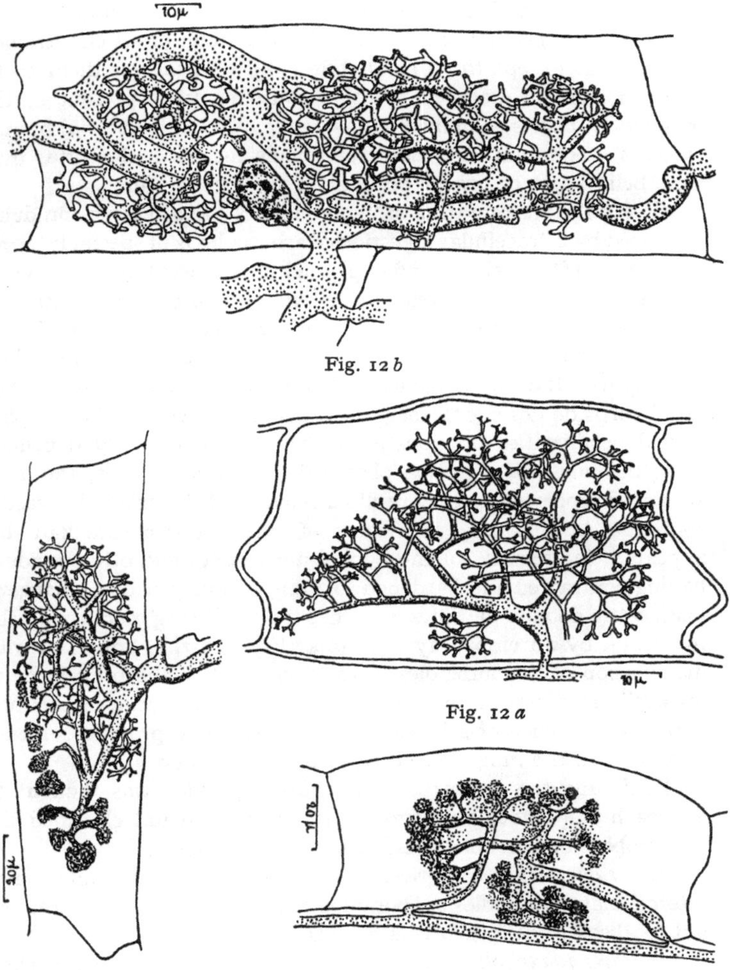

Fig. 12 *b*

Fig. 12 *a*

Fig. 13 *a* Fig. 13 *b*

Figs. 12 and 13. ENDOTROPHIC MYCORRHIZA.

12 *a*. ***Arum maculatum***. Young arbuscule; simple type.

12 *b*. ***Sequoia gigantea***. Young arbuscule; compound type.

13 *a*. ***Allium sphaerocephalum***. Arbuscule in course of transformation to sporangioles.

13 *b*. ***Ornithogalum umbellatum***. Arbuscule undergoing change to sporangioles.

(All after Gallaud, *Rev. Gén. de Bot.* 1905.)

From his own observations he concluded that the production of vesicles was a normal phenomenon for the fungi of all endotrophic mycorrhizas except those of Orchids, and he believed them to be of the nature of storage organs. Arbuscules were present in the majority of plants examined and were regarded as haustorial branch systems functioning as absorptive organs for the endophyte; they were believed to be organs peculiar to mycorrhizal fungi.

Gallaud's interpretation of "sporangioles" as alteration products produced by intracellular digestion of arbuscules is of special interest, inasmuch as it correlated and explained the observations of previous workers and provided a reasonable explanation of the structure of endotrophic mycorrhiza in general. The evidence offered is convincing for the correctness of this view, as is that which established their identity with the "sporangioles," "Klumpen," "masses granuleuses," and "corps de dégénerescence" of previous observers. The morphological peculiarities previously ascribed to them clearly depended upon the phase of digestion observed in individual cells, the most advanced representing only residual remains of the original mycelial contents. At this stage all traces of mycelial structure had disappeared, the remains of the hyphal membranes and other residues having contracted and hardened to form a homogeneous refringent mass with definite outlines attached to the living portion of the mycelium by an empty hyphal stalk (Figs. 13 *a*, *b*, 16). The wide distribution of "sporangioles" was a measure of the prevalence of intracellular digestion in endotrophic mycorrhizas and, in common with the existence of arbuscules, was believed to provide evidence of affinity in the fungi concerned in their formation.

Excluding the mycorrhiza of Ericaceae, which was thought to approach the ectotrophic type in structure, Gallaud classified the endotrophic mycorrhizas studied by him as follows:

1. *Arum maculatum* series. Mycelium at first intracellular, later intercellular; arbuscules or sporangioles usually simple and terminal, not localised in definite layers of cells.

2. *Paris quadrifolia* series. Mycelium always intracellular; arbuscules or sporangioles usually compound, not terminal, localised in definite layers.

3. *Hepatic series.* Mycelium always intracellular with arbuscules and sporangioles not localised in layers.

4. *Orchid series.* Mycelium always intracellular with differentiation to "Pilzwirthzellen" and "Verdauungszellen."

In each of these series Gallaud included plants of the most diverse affinities and habitats. For example, with *Arum maculatum* are grouped Monocotyledons and herbaceous Dicotyledons of many different families and likewise the marattiaceous fern, *Angiopteris durvilleana.* With *Paris quadrifolia*, a like heterogeneous collection of Monocotyledons and Dicotyledons together with *Sequoia gigantea*, and other Conifers, and *Ophioglossum vulgatum.* With *Pellia epiphylla* and other Hepatics were associated the prothalli of *Lycopodium* sp., and with the Orchids such diverse species as *Tamus communis* and *Psilotum triquetrum.*

It is difficult to believe that so heterogeneous a grouping has any real significance. Gallaud himself was careful to point out that his classification was based only on the morphological characters of the mycorrhizas. On critical examination he could find no reasons for believing that the differences described had any taxonomic value in respect to individual endophytes and expressed the view that all the fungi concerned could be placed in a common group:—"Les champignons endotrophes appartiennent à une même groupe dont les subdivisions correspondent aux différentes séries de mycorhizes."

The success attained by a French botanist, Noël Bernard (1903, 1904), in isolating the endophytes of *Cattleya* and other Orchids encouraged Gallaud to make similar efforts on other plants. The fungi subsequently isolated by him, e.g. species of *Fusarium*, *Mortierella*, *Alternaria*, *Cephalosporium*, etc., etc., were correctly regarded as members of the epiphytic fungus flora of the roots. The true endophytes defied his efforts, and he concluded that the intracellular digestion suffered by the mycelium had impaired its power of active growth. His attempts to determine the taxonomic position of the endophytes by inoculation of known forms into seedlings raised under aseptic conditions were equally unsuccessful.

To the solution of the main problem—the biological relation of fungus and host in mycorrhiza—Gallaud's researches contributed little or nothing. Despite the *quasi* parasitic distribution of the intracellular mycelium and his recognition that its harmlessness depended chiefly upon the digestive activity of the root-cells—he anticipated Bernard by comparing these cells with the phagocytes of the animal body—he concluded, in the light of his own observations, that "pour les mycorhizes d'ordre inférieur que j'ai rangées dans les séries de l'*Arum*, du *Paris* et des Hépatiques, il ne saurait y avoir de symbiose harmonique entre la plante et le champignon. Ce dernier est simplement un saprophyte *d'ordre spécial, saprophyte interne dans les racines, où le pouvoir digestif des cellules, sans entraver son développement, empêche qu'il ne leur cause des dommages importants.*" And elsewhere—"On ne peut donc dire qu'il y a symbiose harmonique entre les deux plantes mais bien plutôt *lutte entre le champignon envahissant, mais peu nocif, et les cellules* qui se défendent grâce à leur *puissance digestive.*"

Gallaud's researches followed the conventional lines of investigation laid down by earlier workers. In certain respects, however, his work on mycorrhiza marks the beginning of a new period. More especially in relation to the identification of the endophytic fungi was a changed point of view noticeable: the need for experimental verification was noted, and the absence of any satisfactory proof of the identity of the fungi isolated was frankly admitted. In respect also to the biological relation between vascular plant and fungus, Gallaud showed a disinclination to accept traditional hypotheses, and his own views indicated the necessity for experimental investigation of a more critical kind than had hitherto been bestowed upon mycorrhiza problems.

Although published in 1905, Gallaud's memoir had appeared earlier in the form of a thesis and it is due to this fact that he left unnoticed a paper published at the end of 1904 (Bernard, 1904 *a*) that was destined to bring about a profound change of opinion with regard to the biology of Orchid mycorrhiza.

Two years earlier, in 1902, there had appeared a paper on tuberisation by a French botanist, Noël Bernard, followed in 1903 and 1904 by other communications giving an account of the author's researches on Orchid mycorrhiza. Publication of these papers marks the beginning of a new and fruitful period of investigation on mycorrhiza. In the first, Bernard recorded the isolation of members of the genus *Fusarium* from the roots of various Orchids, believing at that stage of his work that these were the true endophytes. He recorded also the new and remarkable observation that the presence of the endophytic fungi was essential for successful germination of the seeds of various Orchids, and directly related the tuberous habit of the young Orchid embryo with infection by the specific root fungus at germination. He applied the knowledge gained experimentally to explain the development of the tuberous habit in other Orchids, in *Ranunculus ficaria* and in Potato, and elaborated an entirely original theory of tuberisation which will be considered at greater length in a later chapter. Bernard's classical researches on the mycorrhiza of Orchids will now be considered in detail.

CHAPTER V

ORCHIDACEAE

Orchidaceae: Bernard's discovery of the obligate relation in Orchids; the pathogenic theory of infection—Burgeff—Recent researches on the germination of Orchid seeds: Knudson; Bultel; Wolff—The obligate character of infection: Constantin and Dufour; Huber—The symbiotic and asymbiotic methods of germination and their practical applications.

Noël Bernard's experimental studies initiated the modern period of experimental enquiry and mark an epoch in the history of research on mycorrhiza. He approached the subject with a point of view essentially different from that of his predecessors, and, almost at once, reached conclusions respecting the biological relation between fungus and host plant so novel and stimulating that they illumined the subject as a whole and produced a profound effect upon the point of view of subsequent investigators.

To Bernard, the root fungi of the Orchids were parasites, attacking the embryo plantlet at the earliest stage of growth and persisting in the adult as a chronic—although relatively benign—condition of disease. He regarded the Orchids—"comme les plantes atteintes d'une maladie parasitaire chronique qui commence à la germination et persiste en général jusqu'à l'état adulte; maladie bénigne en un certain sens, puisqu'elle n'empêche pas la vie, mais qui ne constitue pas moins une tare physiologique des Orchidées en général, une condition de nature à faire comprendre quelques-unes des anomalies de ces plantes qui passent pour singulière aux yeux mêmes d'observateurs peu initiés à la Biologie végétale." (Bernard, 1904 *a*.)

Bernard's earlier observations were made upon the germination of seeds of *Neottia* and other Orchids, and dealt also with the possibility of correlation between the modes of development of different species and their infestation by the fungal endophytes. A study of the Ophrydeae from this point of view focussed his attention upon the periodicity of tuber-formation in members of this group and its relation to fungal infection of their roots, whence he was led to speculate upon the physiology of tuber-formation in general and to put forward and, subsequently, elaborate an hypothesis to explain the phenomena of tuberisation (Bernard, 1899, 1900 1902 *a*, *b*, *c*, 1903).

This theory, together with its later developments, is dealt with fully elsewhere in the present work, so leaving the way clear for an account of the experimental researches on Orchids and a critical review of the theories based upon them.

As is well known, the seeds of Orchids are extremely small and light and are produced in immense numbers. Darwin (1862) had estimated the number of seeds in a single capsule of some of the common British species at over six thousand and this number is known to be greatly exceeded in some of the tropical species.

At the beginning of the nineteenth century it was commonly believed that they were incapable of germination. Salisbury (1804) was the first to announce that he had observed germination, and he figured the seedlings of *Orchis morio* and *Limodorum verecundum*. The seeds are of very simple structure, each consisting of a small group of similar cells enclosed by a thin reticulate seed-coat (Figs. 14, 18). In certain genera the cells are larger at the suspensor end of the seed; otherwise there is no differentiation of tissue and no distinction into root and shoot such as exists in the embryos of most plants. If removed from the capsule and sown on an ordinary moist substratum away from the parent plant or other Orchids, their behaviour varies with the genus of Orchid. Seeds of *Epidendrum*, for example, undergo practically no change; those of some other genera, e.g. *Odontoglossum*, increase slightly in size and develop chlorophyll; those of *Cattleya* may continue to grow for several months, forming a minute green tuberous structure with rudimentary hairs—the primary tubercle or "protocorm," while in rare cases, as shown by Bernard for the genus *Bletilla*, the seedling may develop a slender stem with foliage leaves. In all cases, if supplied only with the formal conditions for germination, at a critical stage of development varying with the genus of Orchid, growth comes to an end and the seedlings perish. (Pl. II, Figs. 18, 19, 20, 21.)

Bernard's early experience had brought him into touch with the difficulties encountered by practical Orchid growers, many of whom had devised empirical methods to facilitate the successful raising of Orchids from seeds. For example, sowings made upon moss laid upon the pans containing the parent plants, special attention being paid also to conditions such as temperature and moisture. In spite of every precaution suggested by practical experience, the raising of Orchids from seed continued to be a somewhat precarious adventure, leaving growers quite in the dark as to the real nature of the critical factors which controlled germination. Practical experience had shown also that the difficulty of raising seedlings varied greatly with the genus and species of Orchid. Seeds of *Cattleya* and *Cypripedium*, for example, were relatively easy to germinate successfully, while those of *Odontoglossum*, *Phalaenopsis* and *Vanda* presented the greatest difficulty. The irregularities of behaviour observed in horticultural practice were confirmed by Bernard's earlier experiments, which, moreover, failed to provide any consistent explanation of the inconstant results obtained. He noted, however, that the methods of cultivation used in the greenhouses of successful growers must involve the presence of the root fungi in the soil about the roots and eventually their acclimatisation in the plant houses. Hence, he argued, an explanation of the fact that germination of seed, at one time believed to be impossible, had been made practicable by modern conditions of culture, and a justification of the popular method of sowing seed in close proximity to the parent plants. The fact that he had failed to secure, under aseptic conditions, germination of seeds of various Orchids commonly raised by growers, in conjunction with the nature of the successful methods adopted by the latter, had indeed convinced Bernard that the presence of the endophytic mycelium was essential for successful germination, and it became his immediate concern to obtain experimental support for this view.

In the first comprehensive account of his work, published in 1904, his objectives were stated as follows:

1. To germinate Orchid seeds under strictly aseptic conditions on sterilised media suitable for the cultivation of the plants.

2. To isolate and identify the endophytic fungi in pure cultures.

3. To compare the behaviour of infected and uninfected seeds for each species of Orchid.

Aseptic seeds were obtained by careful selection of ripe capsules before dehiscence, external sterilisation of the fruits by means of formalin, and transference of the seeds with aseptic precautions to tubes containing gelatine slopes of a decoction of salep[1].

The record of Bernard's first successful attempts to isolate and establish the identity of the Orchid endophytes has historical interest and merits a brief description. The first was made with seeds of a hybrid *Cattleya* removed aseptically from a ripe fruit and placed to germinate, some under aseptic conditions in the laboratory, the remainder in the greenhouse of a professional grower as in ordinary

[1] The "salep" of commerce is obtained by reducing to powder the dried tubers of various oriental species of Ophrydeae.

horticultural practice. Two months later the aseptic seed cultures were almost unchanged, while those in the greenhouse were germinating and the seedlings showed active fungal infection of the suspensor region. Some of the infected seedlings were transferred to tubes of gelatine media and gave a mixed growth of mycelium, the fungi responsible for which were individually isolated, and tested by inoculation into the aseptic seed cultures. One of the forms thus secured at once provoked germination of a normal and regular kind and was observed to cause characteristic fungus infection, thus providing the necessary proof of its identity with the endophyte.

For the second attempt the fruit of a hybrid *Cypripedium* was available. In this case the endophyte was isolated, not from seedlings, but from fragments of the root of one of the parents (*C. insigne*), the fungal growth from which yielded a form identical morphologically with that obtained from *Cattleya*, and correspondingly effective in provoking the germination of aseptically sown seeds of *Cypripedium*. On the other hand, attempts to establish the identity of a fungus isolated from the roots of *Spiranthes autumnalis* by inoculation into seed cultures of the same species were not successful, thus indicating that the presence of the specific endophyte was not the only critical factor in successful germination.

Bernard was impressed by the difficulty he experienced in isolating the true endophytes from pieces of heavily infected tissues, attributing the relative infrequency of mycelial growth from within to the efficiency of the intracellular digesting mechanism—"en général l'endophyte n'est en voie de croissance active que dans un petit nombre de cellules aux limites des régions infectées. . . ."

The two successful isolations just described are notable as the first in which the identity of a root-fungus was definitely proved by inoculation into seedlings raised under aseptic conditions. They are also of interest as the first experiments providing proof of a definite physiological relation between the endophytes and their vascular hosts, a relation predicted by Bernard in an earlier paper published in 1902.

The fungi isolated from *Cattleya* and *Cypripedium* formed spore-like structures freely in culture and were referred provisionally to the genus *Oospora*, Bernard's earlier attribution of the Orchid endophytes to the genus *Fusarium* being withdrawn as an error of observation due to the prevalence of the members of the latter genus in the epiphytic flora of the roots, and the relatively slight activity of the true endophyte in isolation cultures.

The behaviour of seeds of *Cypripedium* under controlled conditions in the presence of the endophyte was described by Bernard as follows.

Germination was initiated by penetration of the large cells of the suspensor end of the seed, whence the mycelium spread from cell to cell of the middle and outer parts of the embryo which increased rapidly in size and ruptured the seed-coat. At this stage the embryo of *Cypripedium* resembled that of *Neottia* in respect to the precocious appearance of abundant starch in the tissues, the lack of chlorophyll and the absence of root-hairs. Within the inner zone of infected tissue, the cells soon showed the cytological features characteristic of digestive activity, i.e. clumping of the hyphae, hypertrophy of the nuclei, and the eventual appearance of a structureless mass of mycelial residues. In the outer zone of tissue, the mycelium continued to invade fresh cells. With the spread of infection, starch disappeared from the invaded cells, and a terminal meristem, separated from the infected region by a zone of active growth, was formed.

A second stage in development was characterised by the formation of chlorophyll and development of root-hairs involving a sharper delimitation of the infected area. The growing axis remained immune from attack, the mycelium underwent digestion in all the cells of the primary tubercle, and the zone of active infection became restricted to a region between the original point of infection and the apical meristem. The close of this period in seedlings about four months old was marked by the production of the first root. It is of interest to note that the young roots suffered infection, not from the mycelium-filled cells of the primary tubercle, but directly from the soil after emergence. The final stage of germination was marked by a shrinkage of the infected area and elongation of the axis with the formation of leaves and roots, following upon which the seedling took on the appearance and structure of the adult plant.

When cultured with their respective endophytes, the seeds of various other Orchid genera behaved in a manner more or less similar to that just described for *Cypripedium* (Fig. 14). Those of *Bletilla hyacinthina*, an Orchid nearly related to *Cattleya*, responded very differently. The embryo of this species at first enjoyed almost complete immunity from infection and in aseptic culture developed independently to a small plantlet consisting of a slender axis with absorbing hairs and a few green leaves. There ensued a critical stage during which immunity disappeared and, subject to infection by the

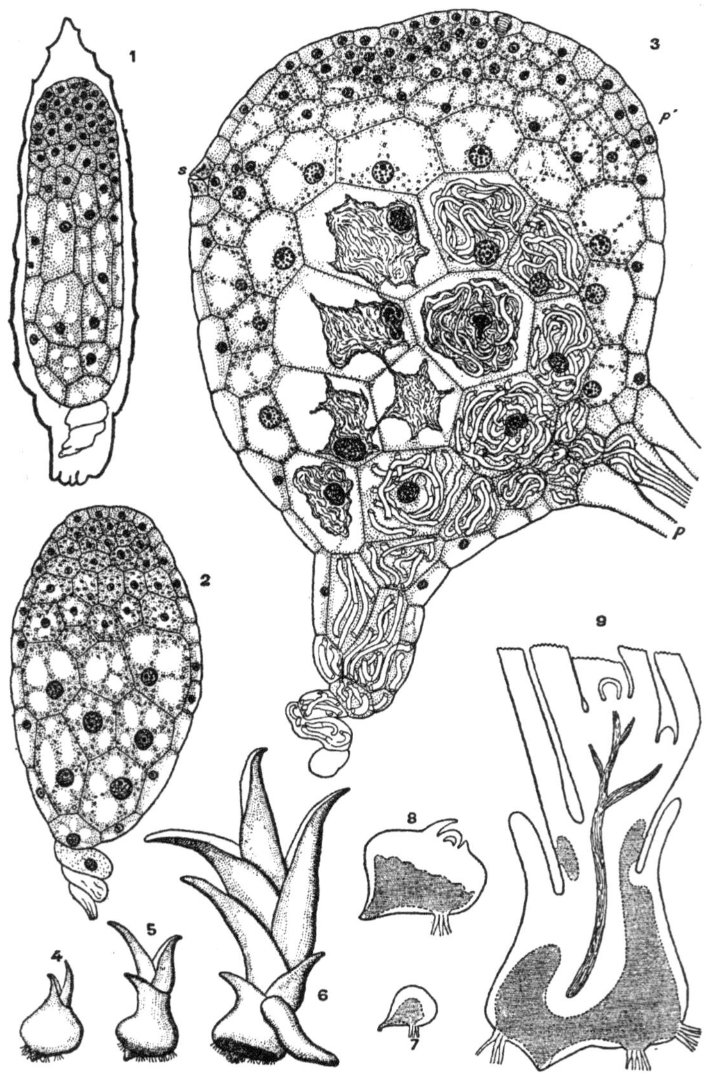

FIG. 14. Seed and Seedlings of *Odontoglossum* hybrids: 1, median longitudinal section of ripe seed showing seed-coat, embryo and remains of suspensor; 2, median longitudinal section of embryo four months after sowing, without fungus; 3, median longitudinal section of seedling, one month after infection by *Rhizoctonia lanuginosa*, *s*, stoma, *p*, absorbing hairs, *p'*, cells divided tangentially, from outer of which absorbing hairs will develop; 4, 3 months' seedling; 5, 5 months' seedling; 6, 7 months' seedling; 7, 8, 9, median longitudinal sections through seedlings of various ages to show extent of infected region in each case (infected tissue shaded). (From Bernard, *Ann. d. Sc. Nat. Bot.*, 1909.)

appropriate fungus, development proceeded normally. Failing such infection, growth came to an end.

Bernard's observations on the cytology of digestion in the embryos and roots of Orchids agreed with those of Magnus and Shibata, but his interpretation of the significance of the changes which took place differed remarkably from that of these observers. Sooner or later, the cytoplasm and skeins of mycelium in the "Verdauungszellen" became reduced to amorphous masses of degenerating material enveloping the nuclei. The nuclear changes accompanying this process, regarded by Magnus and his followers as signs of increased activity, were interpreted by Bernard as signs of degeneration. According to the latter, the infected cells formed, at this stage, a dead tissue:—"Dans la lutte entre le parasite et la cellule qu'il envahit, les deux adversaires ont fini par succomber"!

To summarise briefly the important facts put on record by Bernard in these earlier observations.

Fungal species of similar morphological character were isolated from Orchids belonging to the genera *Cypripedium*, *Cattleya* and *Spiranthes*. Seeds of species of *Cypripedium*, *Cattleya*, *Laelia*, *Brassavola*, and *Bletia*, sown under strictly aseptic conditions, gave normal germination, the seedlings showing typical infection when mycelium of one or other of the endophytes was added to the cultures. Normal growth and development of the seedlings followed upon invasion of certain tissues of the embryos; in the absence of infection, either the seeds did not develop at all, or they did so very imperfectly.

The cases observed were grouped by Bernard in three categories. In the first was placed the genus *Cypripedium*, the seeds of which did not develop at all under aseptic conditions; in the second, *Cattleya*, seeds of which survived for several months in the absence of infection, increased considerably in size, showing a certain degree of tissue differentiation, but, lacking infection, developed no further; finally, there was the interesting case of *Bletilla hyacinthina*, a species with relatively highly differentiated seeds, representing, in Bernard's view, a condition rare among Orchids. From this preliminary work it was concluded that the effect of infection was to stimulate growth, the stimulating effect being invariably manifested in cells remote from the point of infection.

Subsequent observations have shown that, while in experimental cultures the addition of the appropriate mycelium may be delayed until the critical stage in development, under natural conditions infection of the embryo probably takes place much earlier (Fig. 19). In a majority of Orchids, the spread of infection in the embryo is strictly localised, the apical meristem remaining immune to invasion, and the roots showing complete freedom from fungal infection until they reach the soil (Figs. 19, 20, 21). A similar temporary immunity is enjoyed by the annual roots of the tuber-forming terrestrial Orchids, e.g. members of the Ophrydeae. In *Neottia*, the non-chlorophyllous habit is associated with heavier infection; mycelium is more profusely distributed in the embryo and becomes continuous throughout the vegetative tissues of the mature plant. Bernard found seeds of *Neottia* germinating in nature and observed that "elles étaient toutes plus largement infestées encore que les germinations d'Orchidées que j'ai vues dans les serres." He made also the interesting observation that the inflorescences of *Neottia* were sometimes unable to pierce through the overlying humus in which the plant grows. Under such circumstances mycelium spread from the rhizome through the tissues of the inflorescence axis and infected the seeds as they germinated *in situ*.

The isolation of a number of fresh endophytes, with continued study of their characters and relations with different Orchid species, together with the publication of his results, occupied Bernard during the years 1904–1906 (Bernard 1904 *b*, 1905, 1906).

He obtained experimental demonstration of the existence and maintenance of a remarkable reciprocal balance between fungus and Orchid plant. Using seeds of the same Orchid species and mycelium from the endophytes of three different genera, including the parent species, evidence of a surprising degree of specificity on the part of the latter was obtained as follows.

Germinating seeds of a species of *Phalaenopsis* were inoculated from pure cultures of the fungi isolated from *Cattleya*, *Phalaenopsis* and *Odontoglossum* respectively.

In the case of the first the mycelium invaded the seed tissues, no intracellular digestion was observed, and the mycelium parasitised and killed the seed without effecting germination. With the fungus from *Phalaenopsis*, infection took place in the usual way, and was kept within bounds by the digestive activity of the cells of the embryo; normal germination of the type described for *Cypripedium* followed upon infection. Using the fungus from *Odontoglossum*, the intracellular digestion subsequent to infection was excessive, and germination stopped short at a stage marked by the development of root-hairs from the rudimentary axis.

Interpreting these facts in terms of his own theory of infection as a pathogenic condition of relatively slight importance to the Orchid plant, Bernard observed: "à un point de vue théorique, il résulte de ces constatations que l'état dit de symbiose est en quelque sorte un état de maladie grave et prolongée, intermédiaire entre l'état des plantes atteintes d'une maladie rapidement mortelle et celui des plantes qui jouissent d'une immunité complète."

During the same period he perfected a refined and beautiful technique for isolation of the endophytes by direct removal of a young mycelial complex of hyphae from cell to culture dish, thus eliminating the tedious and unsatisfactory procedure involved by isolation of the endophytic mycelium from a mixed growth of fungi associated with the outside of the roots (Bernard, 1909 *a*). He learned also that the endophytes of different Orchids exhibited a well-marked specificity in relation to their individual hosts and further, that the application of his experimental knowledge to the practical problems of Orchid raising was not without its special pitfalls.

Bernard alluded to these difficulties in an address delivered during an International Horticultural Exhibition at Ghent (Bernard, 1908). Three years previously, he had, by request, supplied certain French Orchid growers with *old* cultures of the endophytes from *Cattleya* and *Cypripedium*. The use of this mycelium for inoculation into seed cultures was unsuccessful and gave lamentable results, an observation confirmed by Bernard himself in the laboratory. The reason for this unexpected failure he believed to be due to the use of fungus cultures, which, by long cultivation outside their respective hosts, had lost their power of inducing germination: they were compared directly with the "attenuated strains" of pathogenic bacteria used by Pasteur in the preparation of curative vaccines. Bernard proceeded to confirm this view experimentally by similar methods to those used by bacteriologists, i.e. by re-cultivation of the "attenuated forms" within their specific host plants, and obtained results which led to the belief that complete restoration of the stimulating effect upon germination depended upon the duration of this "restorative" period in relation to the degree of attenuation existing in the first instance.

In the same address, it was suggested that the addition of fragments of the parent roots to the seed bed might faciliate germination under horticultural conditions, the attention of his audience being drawn to the importance of selecting roots containing active mycelium.

Bernard realised that the full fruits of his discovery could only be made available to practical growers by the provision of a specially equipped laboratory, with a constant supply of infected seed cultures in full development upon which to draw for suitable strains of the fungus. In the address just mentioned he observed: "J'ai la conviction profonde que les laboratoires horticoles existiront un jour, mais la symbiose de la science pure et d'horticulture pratique est trop encore dans l'enfance pour que leur utilité s'impose aux esprits."

As he extended his acquaintance with the Orchid endophytes, he modified his earlier views respecting their systematic position. Finally, in view of the marked resemblance of the mycelium to that of *Rhizoctonia violacea*, a fungus common on Potato, he placed them definitely in the genus *Rhizoctonia*. Although impressed by the general similarity of all the forms extracted—"ces champignons sont moins variés que les plantes desquelles ils proviennent"—he gave specific rank to three forms differing in respect to habit and minor morphological details; namely, *Rhizoctonia repens* isolated from *Laelio-Cattleya, Cypripedium* and a number of other genera; *Rhizoctonia mucoroïdes*, from *Phalaenopsis* and *Vanda*; and *Rhizoctonia lanuginosa* from one species of *Odontoglossum*. The forms included under *Rhizoctonia repens* were obtained from many common European Orchids, and proved also to be relatively the most abundant in cultivated species. It may be inferred, therefore, that these fungi have a very wide geographical distribution in nature. All the endophytes doubtless occur in the soil or detritus about the roots of Orchids, but there is at present no exact information as to the occurrence of fungi belonging to the *Rhizoctonia* group in soil.

The mycelium of the endophytes showed a characteristic mode of growth within the tissues, spreading from cell to cell and forming in each a tangled skein of twisted filaments. In Bernard's cultures, a similar tendency to form skeins or "pelotons" was observable, but the habit was not regarded as specially characteristic since it had been observed also in other members of the Mucedineae, the group in which *Rhizoctonia* was included.

When associated with the Orchid plants, spores or other reproductive structures were not produced by these endophytic fungi, but in old cultures the mycelium gave rise to filaments of short segments resembling conidia, and in certain forms these branched hyphae anastomosed to form small yellow or brownish sclerotia, the production of which is characteristic of the genus *Rhizoctonia* as of many other fungi (Fig. 15).

A few words may be added to the accounts already given describing infection in the roots and rhizomes of Orchids. In general, the distribution of mycelium is constant for any one species, but varies in different genera. Hyphae usually enter through the root-hairs but may invade cells of the piliferous layer directly. They pass through the "passage cells" of the exodermis, when this layer is present, and traverse the outer tissues of the cortex in a more or less radial direction—in one species, *Lecanorchis javanica*, Janse described the infecting hyphae as fusing to form a flat mycelial ribbon—and eventually give rise to a zone of typical infection by the formation

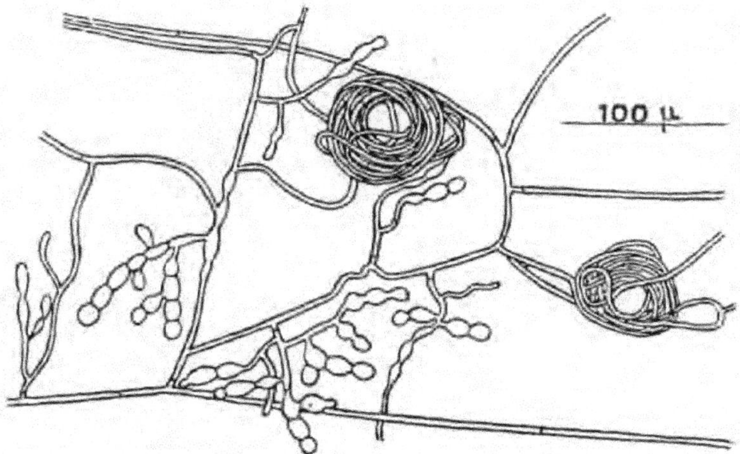

FIG. 15. *Rhizoctonia repens*: mycelium from pure culture. Note chains of conidia-like segments and "pelotons." (From Bernard, *Ann. d. Sc. Nat. Bot.*, 1909.)

of the characteristic skeins or "pelotons" within the cells. The exact position and extent of this infected tissue vary in different genera; in some it is restricted to two or three layers of cells in the middle region of the cortex, in others it occupies almost the whole of the cortical tissue. As infection is localised in the embryo, so in roots and elsewhere it is restricted to the cortex and is absent from meristematic tissues and from those of the vascular cylinder enclosed by the endodermis. In some cases mycelium is confined to localised areas of the root, and occasionally roots may be found entirely free from mycelium, especially in certain genera. Like other chlorophyll-containing organs, aerial roots are usually immune to infection but they may suffer invasion if in contact with the soil and lacking chlorophyll. In the case of individual cells, immunity is indicated, not only by the possession of chlorophyll but also by the presence of special contents —raphides and other crystals, tannin and the like.

The rhizomes of non-chlorophyllous species and of certain green forms show a distribution of mycelium similar to that in roots, and, in general, the tissues of these plants are subject to heavier infection.

Intracellular mycelium in the active condition can best be observed in the tissues nearest to the growing regions; in older parts a majority of infected cells show the mycelium in the "clumped" condition or in various stages of disintegration. As already noted, Magnus described a differentiation of infected cells as "Verdauungs-zellen" (digesting cells) in which the hyphae were destroyed by the digestive activity of the host cells, and "Pilzwirthzellen" (host cells) in which they persisted, retaining their vitality throughout the resting period and serving as centres for re-infection when vegetative activity was resumed (Pl. III, Figs. 22, 23). Magnus attached great importance to this differentiation and to the arrangement in space of the two kinds of cells in *Neottia*, and even suggested the possibility of a classification of Orchids based upon these characters.

Subsequent work has rendered it difficult to maintain the belief that such a regular arrangement exists, even in *Neottia*, and has proved that "host cells" are not present in many genera of Orchids, all those infected being subject to eventual digestion. It is probably safe to assume that the characters shown by the mycelium in the so-called "host cells" represent a relatively early stage of infection preceding the onset of digestive activity.

Certain isolated observations of physiological interest were recorded by Bernard. One related to the marked ability of the endophyte to digest cellulose, another, made several years later, to the fungicidal action exerted by tissue extracts of the tubers of members of the Ophrydeae, these organs of the plants being ordinarily immune to infection (see p. 78).

Of still greater interest was the observation that the stimulus to germination normally provided by the endophytes could be replaced by raising the concentration of organic substances in the substratum used for aseptic seed cultures, i.e. as pointed out by Bernard, by a procedure directly comparable with that known to provoke the development of virgin eggs: "L'action des champignons sur les embryons d'Orchidées peut sans doute être comparée à l'action des spermatozoïdes sur les œufs" (Bernard, 1909 *b*).

These experimental researches on Orchids not only yielded a rich harvest of new and proved facts, but they drew attention to the necessity and value of exact experimental methods in this field of work, and brought the technique of the bacteriologist to the service of the worker at mycorrhiza problems.

As experimental data accumulated, the biological aspects of the symbiotic relation in Orchids and their relation with the phenomena of parasitism, immunity, and pathogenic infection tended more and more to monopolise Bernard's attention. In his most comprehensive work, *L'Évolution dans la Symbiose* (1909 *a*), he elaborated his theory of pathogenic infection, illustrating and supporting his argument by reference to his own experimental results. Another paper dealing with the phenomena of immunity in relation to symbiosis was published in the same year (1909 *b*).

Space does not permit of more than a brief reference to the many interesting facts and reflections brought together in these papers, in which the author sought more especially to study the evolution of the symbiotic habit in plants, and the conditions regulating the "balance of power" between the constituent organisms.

Reviewing the remarkable phenomena of development subsequent to seed infection and noting the obligate character of the association in Orchids, he expressed the opinion that the mutual adaptation of Orchid plant and fungus was a phenomenon linked up with that of the origin of species. Comparative study of the developmental behaviour of different species suggested that the direct ancestors of Orchids were herbaceous perennials with normal germination, the formation of an embryonic tubercle—"protocorme"—having been acquired as a direct consequence of fungal infection. Incidentally, Bernard attached great theoretical importance to the formation of this embryonic structure, in view of its resemblance to the juvenile stage of species of *Lycopodium* and other Pteridophytes.

On the other hand, he held that the relation of germination to fungal infection was not a consequence peculiar to the latter, but could be produced with equal efficacy by the action of concentrated organic solutions. His own experiments had shown—and in view of modern work upon non-symbiotic germination of Orchids it is worth recalling this—that it was easier and more certain, under experimental conditions, to effect the germination of certain Orchids by the action of concentrated organic solutions than by that of the specific endophytes, of which it was often difficult to secure strains in a sufficiently active condition.

Discussing the significance of symbiosis in relation to the evolutionary history of Orchids, Bernard emphasised the comparative ease with which certain species could be raised from seed by asymbiotic methods: "il a été relativement facile de faire germer les Orchidées sans champignon, et il semble bien qu'il faudrait seulement du temps et quelques soins pour prolonger leur culture dans ces conditions" (Bernard, 1909 *a*).

Whatever might have been the significance of symbiosis in the evolutionary history of plants other than Orchids, it was held by Bernard that the habit had played an important part in that of the latter group. The exceptional features shown by *Bletilla hyacinthina* and its near relatives he regarded as rarely preserved characters, representing the nearest approach to the ancestral condition and the starting point of an evolutionary series among living Orchids, of which the last terms were represented by *Cypripedium* and similar types in which the earliest stages of development had become directly dependent upon the nutritive stimulus ordinarily supplied by symbiosis with the appropriate endophyte. In his view, the Orchids represented a remarkable case of adaptation to parasitic invasion, their survival as a group pointing to the possible disappearance of other forms, both plant and animal, which had succumbed to similar attack.

Bernard's later researches were carried out at the Pasteur Institute, a circumstance possibly not entirely unconnected with the fact that he became more and more impressed by the close analogy between symbiotic phenomena and those revealed by the work of Pasteur and Metchnikoff on pathogenic infection: "Dans une étude de la symbiose les expériences de Pasteur doivent servir à éclairer les théories de Lamarck et de Darwin" (Bernard, 1909 *a*). To him the relationship in Orchids represented a condition on the threshold of disease offering an exceptional opportunity for the study of vegetable pathology. The digesting cells of the roots and embryos were directly comparable with the *phagocytes* of the animal body, and the process of digestion with a *phagocytosis* which constituted the chief weapon of the Orchid plant against parasitic invasion of its tissues.

The views expressed by Frank, Shibata and others, who had seen in the phenomena of digestion in the root-cells a mechanism by which mycorrhiza plants obtained a supply of nitrogenous food material from their invaders, were not acceptable to Bernard. This "curious hypothesis," as he described it, was not in his view supported by the observed facts, and was indeed improbable in view of the reactions shown by the cells of Orchid embryos to infection, whether in a condition of balanced symbiosis with the appropriate fungus, or, when suffering damage from invasion by a too virulent strain from another species of Orchid. The formation of special haustorial branches or "arbuscules" by many endotrophic fungi and the conversion of these into structureless "sporangioles" was held to point to a similar conclusion.

Comparison with the mechanism of resistance and immunity in animals was pressed to its limits by Bernard. For example, he claimed that seed cultures inoculated by a strain of *Rhizoctonia* too weak to effect germination, became incapable of development when subjected to the action of a suitable strain—they had been "vaccinated" and so rendered immune to the action of the more "virulent" form—a strange reversal of the result desired in ordinary vaccination!

The *obligate* condition in Orchids was held to be associated directly with a secondary effect of infection, viz. a rise in the concentration of the cell sap—in his view, equally well brought about under experimental conditions by chemical means:—"Le phénomène de croissance est la manifestation visible de transformation du contenu de la cellule ayant pour résultat d'augmenter la quantité absolue de substances dissoutes dans le suc cellulaire" (Bernard, 1904 *a*). While he admitted the possibility of autonomous existence for certain species in the adult state, Bernard knew of no case of Orchid germination lacking fungus infection in nature or horticultural practice, although such had been observed experimentally.

Discussing the possible existence of species immune to infection in the adult state, he recalled Frank's observations claiming immunity for roots of *Epipactis latifolia* and *Listera ovata*, the view of Johow respecting *Wullschlaegelia* and that of MacDougal for *Cephalanthera oregana*. A similar temporary freedom from infection in mycotrophic plants other than Orchids had been noted by himself in *Psilotum triquetrum*, and by Gallaud (1905) in *Ranunculus ficaria* and *Arum maculatum*. In none of these cases, it was pointed out, had any experimental evidence been offered that *germination* and *development* could take place independently of infection. From his own observations on *Epipactis*, he thought it unlikely that such would be forthcoming. Of interest in the same connection may be noted again the partial and intermittent immunity shown by a majority of Orchids, as exemplified by the complete freedom from infection of certain tissues, e.g. meristems, chlorophyllous cells and the tissues of the vascular cylinders, and by the progressive localisation of infection in the tissues of germinating embryos. Moreover, in some species, e.g. *Neottia*, fungal infection was found to be extensive and continuous throughout life; in others, e.g. members of the Ophrydeae, it was periodic and intermittent, the roots of each season's growth being newly infected from the soil after emergence. Bernard was concerned to find a physiological interpretation of these facts and, in a posthumous paper, described the fungicidal effects observed in the tuber tissues of members of the Ophrydeae (Bernard, 1911 *b*).

His early observations on this group of Orchids had shown that the tubers were immune from infection. Experimentally, it was found that small pieces of tissue, removed aseptically from a living tuber and imbedded in culture vessels of gelatine nutrient inoculated with one or other of the endophytes, caused arrest of growth and death of the protoplasm, as the mycelium came within range of substances diffusing from the fragments of tuber tissue. The fungicidal effect was produced only by living tissue, the lethal substance being destroyed by temperatures above 55° C. Extending these observations, he reached the conclusion that there was produced, in tubers of the Ophrydeae, a fungicidal substance, easily diffusible and readily destroyed by heating, the action of which was specific upon the endophytes and varied with the forms extracted from different species of Orchid.

Bernard's results have since been confirmed by Nobécourt (1923) and by Magrou (1924 *b*). The former found that the action of the tuber was inhibited, not only by heating, but also by chloroform and by cooling to 15° C. He believed that the fungicidal substance was a secondary product formed by the interaction of toxins from the mycelium with substances diffusing from the tuber tissue. Bernard had been prepared to regard the effect as due to an "antibody" secreted by the intact tuber under the influence of an "antigen" diffusing from the mycelium present in the roots. Magrou's recent experiments afford confirmation of the latter view by showing that diffusion from the fragment of tuber, before inoculation by the fungus, is sufficient to render the gelatine medium toxic to mycelial growth.

Noël Bernard's brilliant researches were brought to a premature end by his untimely death in 1911. He had initiated a new line of experimental investigation and made it peculiarly his own. Already, in one short decade, he had changed the whole aspect of the mycorrhiza problem, bringing it into touch, on the one side with the physiology of parasitism and pathogenic infection, and, on the other, with the difficulties experienced by practical horticulturists. The knowledge acquired by experiment in the laboratory was placed at the service of Orchid growers and, in the hands of some of them, has since proved to be of great practical value. He left to his colleagues and successors the task of completing his work; in particular, of providing the requisite experimental support for his novel and stimulating theory of tuberisation.

His researches had already given an impetus to research on mycorrhiza, and the success of his experimental work on Orchids was doubtless indirectly responsible for the publication of a monograph entitled, *Die Wurzelpilze der Orchideen*, in which Burgeff (1909), a German worker, contributed an account of his own experiments, confirming many of Bernard's results and extending the latter's observations in certain directions.

Burgeff expressed the view that the endophytes belonged to a single group of Fungi, the Orcheomycetes, for which he proposed the generic name *Orcheomyces*, without reference to the systematic affinities of the genus. He added to the list of root fungi already isolated by Bernard and classified the known forms in five groups, each including a number of forms to which he gave specific rank; he contributed also additional details as to the specificity of forms isolated from particular Orchid species. Devoting special attention to the physiology of the root fungi in pure culture, he put on record a number of new observations bearing on their metabolism, of which perhaps the most significant related to their inability to grow on substrata lacking combined nitrogen, and hence, to the absence of any positive evidence for the fixation of atmospheric nitrogen. In view of the recent researches of Wolff (1925) on the endophyte of *Neottia*, it is clear that this conclusion may require revision (see p. 92).

Burgeff was unable to confirm Bernard's observation that the efficacy of the endophytes in respect to germination was impaired by prolonged cultivation outside their host plants, finding, on the contrary, that a culture 26 months old showed no decline of activity when used for the inoculation of aseptic seed cultures. He believed the activating effect of the mycelium upon development to be due to the production of enzymes which acted upon the carbohydrates present in the seed tissues and thus expedited germination. Bernard had related the stimulus following infection to a rise in concentration of the cell sap. On Burgeff's hypothesis, this would follow from an increased sugar content, due to the diastatic activity of hyphae in contact with the starch-containing cells of the embryo and roots, thus accounting also for the observed disappearance of starch from these tissues subsequent to infection.

Burgeff was Stahl's assistant, and this monograph on the Orchid fungi was doubtless prompted in part by a desire to relate the new facts established by Bernard to the older theories of nutrition put forward by Stahl and others. On theoretical grounds he found himself in general agreement with Stahl in respect to the existence of a symbiotic relation beneficial to the vascular plant. His researches have contributed many new facts respecting the physiology and nutrition of the Orchid endophytes in pure culture.

To Kusano (1911), a Japanese observer, is due the recognition of an unusual and remarkable type of mycorrhiza in *Gastrodia elata*, a curious non-chlorophyllous Orchid from Japan. The species is described as widely distributed in Japan, occurring chiefly in woods of *Quercus serrata* and *Q. glandulifera*. It had been mentioned earlier by Johow (1889) as a humus saprophyte with mycorrhiza, but, previous to the publication of Kusano's paper, nothing definite was known respecting either the type of mycorrhiza or the mode of nutrition.

The plant appears above ground only in the flowering stage, the vegetative body consisting of a system of subterranean rhizomes and tubers bearing scale leaves. The large flowering tubers are 10 cm. to 17 cm. in length and give rise to lateral offsets in the form of small daughter tubers. At the end of May the flowering tuber produces an immense inflorescence a metre or more in height. Externally, the massive tubers are covered with a corky skin like that of a potato, enclosing a parenchymatous tissue with small centrally placed vascular strands. In the autumn, the parenchyma contains abundant carbohydrate reserves, the chemical nature of which was described in detail by the author.

Kusano's interest in the plant was first aroused by observing the relatively small proportion of flowering tubers as compared with the number of offsets produced, and also by noting the freedom of the latter from any trace of fungus infection. Of mycorrhiza, in the strict sense, there was none, since the plant is rootless.

The offsets showed no signs of fungus infection, but the flowering tubers were found to be invariably associated with the rhizomorphs of *Armillaria mellea*, the Honey Agaric. This fungus, known to foresters in Europe as a most destructive parasite, was described by Kusano as common on the older roots of the two species of Oak above mentioned where it was regarded as growing saprophytically without forming typical mycorrhiza in the younger roots. The matted strands of mycelium or rhizomorphs were common in the soil about the trees, frequently spreading to potato fields in the neighbourhood, in which case the fungus behaved as a true parasite, attacking and destroying the potato tubers. Infection of the *Gastrodia* tuber by the rhizomorph of the fungus was effected by haustorial branches which penetrated the corky covering and formed localised areas of infection, the whole relation showing a close superficial resemblance with that of a *Cuscuta* plant and its host. The infected area of the tuber showed three distinct regions differing in the structure of the cells and the character of the intracellular mycelium. The hyphae in the outer tissues remained unaltered: in the middle region they attacked and consumed the protoplasm and nuclei, passed through a stage that simulated "clumping" in the root cells of Orchids and subsequently underwent complete auto-digestion: in the inner tissues they were subjected to rapid digestion by the cells of the tuber. Remarkable cytological changes in the infected cells were described by Kusano. The cells of the inner region in particular showed signs of great metabolic activity, becoming filled with secondary food products. In the earlier stages of infection starch disappeared from all the infected cells, re-appearing in those of the inner region at the close of the period of active digestion.

The association between tuber and fungus was apparently casual and took place only occasionally. When it occurred, a full-grown offset was formed which flowered the following spring, otherwise small daughter tubers were produced which dwindled without flowering (Pl. III, Figs. 27, 28).

The relation in this species of *Gastrodia* is a very puzzling one. Subsequent to the formation of external cork, the tubers are unable to absorb liquids from the soil and apparently depend solely upon the chance of infection for a means of communication with the outside world. It is not clear that the fungus can derive even a temporary benefit from the association although this is evidently of a parasitic nature in the first instance. On the other hand, the *Gastrodia* plant evidently profits, since the ability to form flowers and fruit depends directly upon invasion. It was suggested by Kusano that the Orchid plant has converted a parasitic attack by the rhizomorphs into a means whereby food materials are transferred from the Oaks or other plants to which the mycelium of the fungus is attached. If so, a remarkable case indeed of a vascular plant turning the tables upon one of the most destructive fungus parasites known!

There is as yet no information as to the behaviour of seeds of *Gastrodia* nor evidence of the existence of any relation with an endophyte of the *Rhizoctonia* type common to other Orchids. At present the case stands as an isolated and unique instance of a non-chlorophyllous vascular plant which depends upon a parasitic relation with a fungus for the completion of its reproductive cycle.

The physiology of another species of *Gastrodia*, *G. sesamoides*, has recently been studied by McLuckie (1924). This endemic Australian Orchid possesses a perennial rhizome with a few roots and colourless scale leaves. The flowering shoot develops from the apex of the rhizome and consists of a brown, slightly succulent axis up to eighteen inches in length bearing a few membranous scale-like leaves and a terminal inflorescence.

The rhizome shows both fungal and bacterial infection. The former is limited to isolated groups of superficial cells whence hyphae extend outwards into the soil; the coarse non-septate mycelium is intracellular in distribution but is not digested by the host cells. Bacterial infection is profuse, affecting not only the tissues of the rhizome but also those of the roots and flowering shoot. In artificial cultures the bacteria were found to assimilate free nitrogen at the rate of 7·8 mg. of nitrogen per 100 c.c. of culture solution in 15 days.

This species of *Gastrodia* had been previously recorded as a root parasite. The observations just recorded led McLuckie to class it as a saprophyte associated symbiotically with a mycorrhizal fungus and a nitrogen-fixing bacterium. It is believed that the Orchid is directly and indirectly dependent upon its endophytes for both carbonaceous and nitrogenous food-materials, and it is suggested by the author that the association is an obligate one. It is thought that the mycelium functions by absorbing water and mineral salts, and is likewise responsible for the intake of organic carbon compounds and possibly also of organic nitrogen compounds from the soil. Assuming the correctness of this interpretation, the bionomics of the species form an interesting contrast with those of *Gastrodia elata*.

Another uncommon non-chlorophyllous Orchid, *Dipodium punctatum*, has also been investigated by McLuckie (1922*a*). This is a terrestrial plant occurring in the "shaded humus soils of the Australian Bush." The reddish-coloured flowering stems, bearing small scale leaves, arise from a few large succulent roots which penetrate the humus in various directions. Like *Monotropa*, the species had been described as a root parasite on neighbouring plants (Moore and Betsche, 1893). No evidence of parasitism was found by McLuckie, whose observations led to the view that in this case also the root fungus supplied its host not only with organic substances from the humus, but also with water and salts. Whence followed the logical conclusion that "in *Dipodium*, 'symbiotic saprophytism' has practically become a case of the higher plant being parasitic upon the endophyte."

The roots are remarkable in the possession of a sheath resembling the velamen of aerial roots, the thin-walled, living cells of which fit closely together and appear to constitute an "aqueous tissue." The mycorrhiza is of the familiar Orchid type with extensive intracellular infection of the cortical cells, followed by digestion and disappearance of the stainable products.

McLuckie's conclusions as to the real significance of root infection in this Orchid are supported by evidence of the usual kind—absence of root hairs and an assimilatory mechanism in the vascular partner, the known ability of the endophyte to utilise organic compounds of carbon and nitrogen, wholesale digestion of the intracellular mycelium—there is, indeed, a strong presumptive case for parasitism on the endophyte, not only in *Dipodium*, but in most of these so-called vascular "saprophytes." The pressing need at the moment is for *experimental* evidence bearing on the physiological relations between the symbionts, and for proof that the vascular plant alone is unable to utilise the nutritive substances locked up in humus. Such evidence can be obtained only by "pure culture" methods, which present exceptional difficulties in the case of non-chlorophyllous species, owing to the initial difficulty of securing seed germination.

Knudson (1922) began a fresh chapter in the story of the Orchids and their root fungi with his paper entitled, *Non-symbiotic germination of Orchid Seeds*, to which two others have since been added (Knudson, 1924, 1925). His interpretation of the observed facts differs from that of Bernard; following a review of the three papers just mentioned the nature of this disagreement will be discussed. In the first, he described the germination of seeds of *Laelia*, *Cattleya* and related forms in aseptic cultures by the addition of certain sugars, or of plant extracts with traces of sugar, to the substrata. He recorded also the interesting new observation that inoculation of the seed cultures with a strain of *Bacillus radicicola* isolated from Alfalfa, produced a beneficial effect upon germination and the development

6-2

of chlorophyll. Two years later he reported that he had obtained practically 100 per cent. germination of seeds and the subsequent production of healthy seedlings of species of *Cattleya*, *Laelia*, *Epidendron*, *Cymbidium*, *Phalaenopsis*, *Dendrobium*, *Ophrys*, *Cypripedium* and *Odontoglossum*, i.e. that these non-symbiotic methods can be used with equal success for raising seedlings of the more refractory species. He found likewise, that when germination had advanced to a certain point, the seedlings became independent of an artificial supply of sugar, and could be transplanted to a suitable nutrient containing inorganic salts only, where they continued to make healthy growth quite independently of fungus infection (Pl. III, Fig. 26).

From these experiments Knudson deduced certain conclusions among which may be noted the following: (1) that the necessity for symbiotic fungus infection has not been proved; (2) that the germination of Orchid seeds is dependent on an outside source of organic matter, e.g. sugar, the effect of which is purely nutritional; (3) that young Orchid embryos, germinated aseptically without sugar, are unable to photosynthesise even if chlorophyll is present, such inability being presumably due to lack of some internal factor.

Knudson subsequently extended his researches and repeated and confirmed certain experimental observations made by Bernard. Thus, he isolated the endophytes from *Cattleya*, *Cypripedium*, and *Epipactis* and used the three strains—apparently regarded by him as identical in physiological action as in morphological character—for the successful germination of *Cattleya* seed. He made two observations of special interest, one, that the inoculation of seed cultures with certain other fungi, e.g. *Phytophthora* sp., produced comparable effects—"*Phytophthora* sp. is about as favourable to germination as the orchid fungus" —a fact already put on record by Burgeff for casual infection by *Penicillium*; the other, that successful non-symbiotic germination was obtained on a sterilised peat and *Sphagnum* mixture by the addition of a solution of inorganic salts of hydrogen-ion concentration equal to *p*H 4·6, the germination being equally rapid as when the fungus was supplied. No explanation whatever of this latter observation can at present be offered beyond the assumption put forward by the author himself; namely, that under the conditions described, the embryos obtained the requisite soluble organic food material direct from the sterilised substratum. It must be noted that the seeds used were those of *Cattleya* which are known to reach a fairly advanced stage of development in aseptic cultures without inoculation

In short, Knudson has confirmed the *fact* of symbiotic germination under experimental conditions but rejects Bernard's views as to its significance. Discussing this disagreement, Knudson stresses the importance of the nutritional aspect of the problem. In his view, the sugar supplied to germinating seeds acts as a food, and the beneficial result of inoculation by a suitable fungus in Bernard's experiments depended, not upon infection of the embryo tissues or upon any internal effect, but upon the production of sugar from organic material supplied in the medium with a corresponding rise in the concentration of the substratum. The effect produced in Knudson's cultures both by the specific endophytes and by certain foreign organisms is similarly explained. In addition to the production of sugar from starch, Knudson noted a marked rise in the hydrogen-ion concentration of inoculated cultures. For example, in one series, concentrations equal to *p*H values of 7·0, 6·3 and 6·0 were changed to *p*H values of 4·6, 4·8 and 4·0 respectively, indicating a decided increase in the acidity of the substratum. In the more acid media healthy seedlings were produced, in the others the embryos perished. Commenting on these facts he observes: "Here is evidence that the hydrogen-ion concentration is markedly increased by the fungus, and this must be due to organic acids produced by the fungus."

Knudson confirmed Bernard's and Burgeff's observations on the specificity of the endophytes but again offered a new interpretation of the observed facts. Noting that seeds of *Odontoglossum* were parasitised and killed by the fungal strains isolated from *Cattleya*, *Epipactis* and *Cypripedium*, he concluded that "these fungi are extremely pathogenic." But selective pathogenicity of this kind at once raises biological problems of profound interest, and it was to the analysis of such problems that Bernard devoted his last years. Had he lived longer, it can hardly be doubted that he would have turned once more to the practical applications of his discoveries, more especially to the improvement of methods for effecting non-symbiotic germination. He had already discovered and noted that this method was safer and easier than that of fungus infection, owing to the difficulty of maintaining suitable strains of the endophytes in artificial cultures.

Knudson's more important theoretical conclusions are indicated in the following passages: "The explanation of the failure of orchid seeds to germinate when provided with all the conditions that permit of the germination of most seeds, is to be found in the organic food relations. The seeds of orchids are lacking in food reserves. . . . Growth of the embryo will continue for a time at the expense of the reserve food, but ceases sooner or later. If the embryos are then supplied with sugar, growth will continue and germination occur. . . . If the embryos are carried over this critical period, then they are thereafter self-sustaining. The significant fact has been noted that the seeds of terrestrial orchids may germinate in nature under conditions where chlorophyll is entirely lacking. These embryos are purely saprophytic. The conclusion seems to be warranted that under natural conditions the orchid embryos are dependent for continued development on an appropriate supply of organic food, which must be absorbed from the material on which the seeds are germinating. Under natural conditions, this food is made available to the orchid embryo by the digestion of organic matter, which transforms the insoluble substances to soluble products. Some of these substances are absorbed by the embryos and are used in the metabolic process. Under natural conditions the orchid fungus may function in these digestive processes, but it would be pure assumption to conclude that no other micro-organisms are involved in this transformation. Sugars are undoubtedly formed, although the concentration would be low. It would seem that other substances are more effective."

The interest and practical value of Knudson's work is not to be denied. To the writer it seems unfortunate that he should have marshalled his experimental results in the form of an argument against the "so-called symbiotic theory of germination" attributed to Bernard and Burgeff. In the main, the new facts constitute an extension rather than a refutation of Bernard's work. It is true that the latter, who originally recorded the rise in concentration of the substratum brought about by fungus growth and also the practicability of non-symbiotic germination, was temporarily too preoccupied with the mechanism of immunity and resistance in Orchids to carry further his observations on the parallel effects produced by infection on the one hand and an increased concentration of sugar *outside* the tissues on the other. It is equally true that the explanation put forward by Knudson, providing as it does an adequate and intelligible physical basis for the effect produced by the presence of the fungus, whether in the tissues of the embryo or in the surrounding medium, does not cover all the known facts. It does not explain, for example, the invariable association of particular fungus strains with individual species of Orchid in nature, the toleration of mycelium in certain tissues, or the fungicidal action exerted by others.

Observations on the effects produced on germination in Bernard's experiments by the use of cultures of different ages led Knudson to conclude that "the most active fungus is the weakest pathogen." They led Bernard, with equal justice, to the view that the absence of obvious pathogenic symptoms in Orchids was due to an acquired resistance similar in kind to that displayed by the "carrier" in the realm of animal pathology. To him, the Orchid fungi were all pathogens. As with pathogenic organisms which attack animals and man, the degree of "virulence" exhibited would be expected to vary directly with the condition of the host and with the history of the attacking organism in pure culture. He was far from underrating the risks involved in the symbiotic method of germination under laboratory conditions. "For five years," he writes, "I have sown seeds of various species of Orchid in culture tubes each containing one hundred seeds, and have inoculated these cultures with *Rhizoctonias* extracted from the roots....On the whole, I have obtained some hundreds of plants, but I underestimate when I place the total number of seeds used in my experiments at 50,000" (Bernard, 1909 *a*).

That healthy seedlings of certain Orchid species can be grown for several years under experimental conditions without infection has been proved by Knudson and others. The problem which engaged Bernard's attention was not the production of such artificially "sheltered" plants, but the explanation of a condition found *invariably* in nature and its relation, if any, with the observed irregularities in seed germination.

In the meantime, any complete explanation of the association of Orchids with particular root fungi must take into account the facts just mentioned. For which reason the "further investigations on the food relationships of orchid embryos" promised by Knudson will be awaited with great interest.

Constantin and Magrou (1922) have subjected the views expressed by Knudson in his earlier paper to severe criticism. It has been pointed out by these authors, as in the present review, that the observation recording the fact that the effect of the fungus on germination could be replaced artificially by a supply of sugar or other nutrient is not a new one, and in itself provides no explanation either of the invariable presence of the endophytes in Orchid roots, or of the specificity exhibited by the forms associated with particular Orchid species.

That root infection is practically invariable in Orchids has been amply demonstrated by Wahrlich and others. For further evidence on this matter attention is directed to the more recent observations by Constantin and Dufour (1920) on *Goodyera repens*. Examination of the rhizomes of an immense number of plants of this species by one of these observers, not only confirmed the fact of invariable fungus infection, but provided evidence of the presence of an identical strain of the endophyte, named by the authors *Rhizoctonia Goodyera repentis*, in every case examined.

The researches of Huber (1921) on *Liparis loeselii* are significant in the same connection.

Since the middle of the nineteenth century when they were described by Irmisch (1847, 1850, 1854, 1863), the European Orchids belonging to the small group Malaxoideae have been known to possess morphological peculiarities, e.g. the presence of aerial tubers resembling those of tropical species. In an account of these special features Goebel (1901) included a brief account of the remarkable distribution of mycelium in the vegetative organs. With this exception nothing was known of the mycorrhizal relations previous to Huber's work, a fact doubtless accounted for by the extreme rarity of the species.

Huber confirmed Goebel's observations and undertook an investigation into the history of infection in the individual plant of *Liparis loeselii* together with the isolation of the endophyte and its behaviour in pure culture. The facts elucidated in the course of this research are of interest and have a direct bearing on the existence or otherwise of an obligate symbiotic relation.

The isolation of the endophyte presented no difficulty. It proved to belong to the type, *Rhizoctonia repens* Bernard, or *Orcheomyces psychodis* Burgeff, and agreed with the forms previously described by the latter author in respect to nutrition. Two features of special interest may be noted in Huber's account; namely, the unusual distribution of fungus infection in the vegetative organs, and the artificial production and attempted cultivation of plants entirely free from fungus infection.

Liparis is a chlorophyllous species with no obvious peculiarities relating to nutrition. The rhizomes show invariable and profuse fungus infection of the cortical tissues, the mycelium in a majority of the infected cells eventually undergoing digestion in the usual way. The roots and leaf-bases are also infected but the amount of mycelium present in these organs is relatively scanty. In the epidermis of the roots and leaf-bases, the endophyte forms chains of spores identical with those observed in pure cultures. Seed germination is replaced by a profuse development of adventitious buds on the aerial tubers. With regard to the course of infection, the mycelium present in the old rhizome does not spread into that of the succeeding year. Infection of the latter is accomplished indirectly through the roots which become invaded by growing into the cortical tissues of the old axis. In the initial stages, therefore, the bud from which will develop a new vegetative axis is entirely free from mycelium, and, if separated with due precautions from the older parts, can be further grown independently on a sterilised substratum.

By taking advantage of this, Huber was enabled to test the capacity of the mature plant for independent growth lacking fungus infection. Buds were removed in January and developed freely to a height of some centimetres on a sterilised substratum. Microscopic examination showed the plantlets to be entirely free from fungus mycelium, the cells of the cortex possessing small nuclei and an abundance of starch and thus differing markedly from those in normally infected plants. Subsequently, growth fell behind that of the uninfected controls; the infected plants became more susceptible to adverse external conditions, did not produce flowers, and by the middle of July all had succumbed. It was inferred by Huber that infection is an obligate condition for full development: "Die Unentbehrlichkeit des Pilzes auch für die erwachsene Pflanze ist damit erwiesen."

Apart from its bearing on the question of non-symbiotic germination, the condition described for *Liparis* shows many features of interest. It has been suggested by Huber that the symbiotic relation in the Malaxoideae is relatively simple, the endophyte showing little modification and the infected tissues being relatively primitive in respect to the differentiation of "Pilzwirthzellen" and "Verdauungszellen."

The lack of agreement between the conclusions of Bernard and Burgeff respecting the efficacy of old fungus cultures to effect germination has been investigated experimentally by J. Wolff (1923), who has expressed the opinion that Bernard did not sufficiently take into account the age and origin of Orchid seed used for experiments in estimating the effect of the age of fungus cultures upon germination. Wolff's experimental results indicate that the rate of germination and the germination capacity of seeds of some species of Orchids decrease rapidly with age, those more than three months old failing to germinate at all. It was recorded subsequently (Wolff, 1925) that of seeds sown on a medium containing 2·5 per cent glucose, some lost their germination capacity in 45 days, others in two to four months; if stored dry, *in vacuo*, they retained their vitality for longer periods. Knudson (1922) also reached the conclusion that Orchid seeds should be germinated as soon as possible after collection.

Wolff (1924) has also investigated the injurious effects sometimes observed when Orchid seeds are germinated in highly concentrated media in contact with mycelium of the appropriate endophyte. By changing the composition of the medium, it was found possible to produce seedlings of *Cattleya* resistant to the injurious effects of a mycelium pathogenic to them under other conditions. For example, seeds of *Cattleya* germinated asymbiotically on a medium containing glucose were able to resist attack by a mycelium which killed seedlings germinated directly in its presence.

The use of "pure-culture" methods for the raising of Orchids from seed has been adopted with success by practical growers both in Great Britain and other countries. An admirable account of the remarkable work accomplished by the late Mr Charlesworth in his own nurseries has been contributed by Ramsbottom (1922), who was associated with the work (Fig. 25). Analogous results have been obtained in Belgium and in Germany. To the success of the method in the hands of French growers witness has been borne by Bultel (1920, 1925) and Constantin and Magrou (1922). (Pl. III, Fig. 25.)

To an account of his own successful experiments on the raising of Orchids, M. Bultel added: "La culture que nous pratiquons de préférence est la culture aseptique en tubes sur milieux gélosés ou autres ensemencés du Champignon endophyte; ce procédé présente sur tous les autres le grand avantage d'une *réussite assurée*, et aussi celui de ne demander aucun soin depuis le jour de semis jusqu'au moment du repiquage des jeunes plantules, soit plusieurs mois après."

More recently the same observer (Bultel, 1925) has reviewed the older work on the symbiotic relations in Orchids, and discussed the degeneration and regeneration of the endophytes in artificial cultures. He has described, also, his own methods for isolating the root fungi and for raising seedlings both by symbiotic and asymbiotic methods. The contention that Orchid seedlings raised asymbiotically grow into abnormal plants incapable of flowering is not supported by his experiments. On the contrary, he has placed on record the case of two hybrid Orchids, raised without infection, that produced flowers in the normal way.

To the professional horticulturalist, the symbiotic method of germination still presents obvious difficulties. The isolation of the specific endophyte is a delicate operation, as is the maintenance of strains in a suitable condition for promoting germination. Moreover, the sterilising of culture vessels and media involves the use of special apparatus not ordinarily needed in horticultural practice. As pointed out by Bernard himself, the use of non-symbiotic methods would remove several of these practical difficulties and greatly expedite the raising and cultivation of Orchids under glass. More especially would this be so, if the claims made recently by Knudson and others in respect to the growth of fungus-free plants stand the test of further experiments and can be shown to apply to horticultural conditions. In the meantime, practical growers in Great Britain and elsewhere are experimenting with non-symbiotic methods, and in certain cases have reported very favourably.

A remarkable collection of seedling cultures raised asymbiotically on artificial media containing sugar and other organic substances was exhibited recently in London, the tubes in every case being crowded with healthy seedlings. The cultures included seedlings of species of the more refractory genera, e.g. *Odontoglossum*. The exact constitution of the media used was not stated, but the worker responsible for these cultures has placed on record certain interesting facts relating to his own experiments. He has confirmed Knudson's observation that the reaction of the substratum exerts an important effect upon germination. (Pl. III, Fig. 24.)

For seeds of *Odontoglossum*, the best results were obtained with *p*H values ranging from 6·5 to 6·8. Special attention has been paid to this genus, and it was found that a medium which promoted germination of seeds of one species of *Odontoglossum* was unsuited for those of another and that it was necessary to carefully adjust the constitution of the media to the needs of individual species. The exact significance of these facts is not at present clear, but they are of interest in view of the difficulties experienced in obtaining successful germination of seeds of *Odontoglossum* by the ordinary horticultural methods (Clement, 1924).

No aspect of the physiology of the symbiotic relation in mycorrhiza is of greater interest than that relating to the possibility of nitrogen fixation by the endophytes. A preliminary account of new researches on the root fungus of *Neottia* by H. Wolff (1925) has reopened this question in relation to the Orchid endophytes.

Wolff reports that the fixation of atmospheric nitrogen by the endophyte of *Neottia* has been quantitatively proved in pure culture, the fungus being able likewise to use organic compounds, e.g. glycocol, and also ammonium salts as sources of nitrogen. A large number of compounds were utilised as sources of carbon, e.g. glucosides (tannin), polysaccharides, hexoses, pentoses and pentosans. Fuchs and Zeigenspeck (1924) had previously shown that the fungus of *Neottia* could utilise similar compounds in nature, and had demonstrated the presence of the appropriate enzymes in infected root cells. It is surmised by Wolff that these enzymes are produced by the mycelium and not by the cells of the host. His present researches seek to demonstrate that compounds of carbon and nitrogen formed by the fungus eventually fall to the share of the *Neottia* plant. More recently (1926) these conclusions have been extended to other species. (See p. 208.)

It is worthy of note that the small family Burmanniaceae, containing a number of remarkable non-chlorophyllous species of very reduced habit, is closely related to Orchidaceae. (See Plate I.) Like the Orchids, members of the former group produce seeds with scanty reserves and also develop typical mycorrhiza in the root system.

There are at present no experimental data concerning germination or the biology of infection and its relation with nutrition, but it would not be surprising to learn that in this group, as in Orchids, there has been evolved an association more intimate than that found in mycorrhizal plants generally.

PLATE II

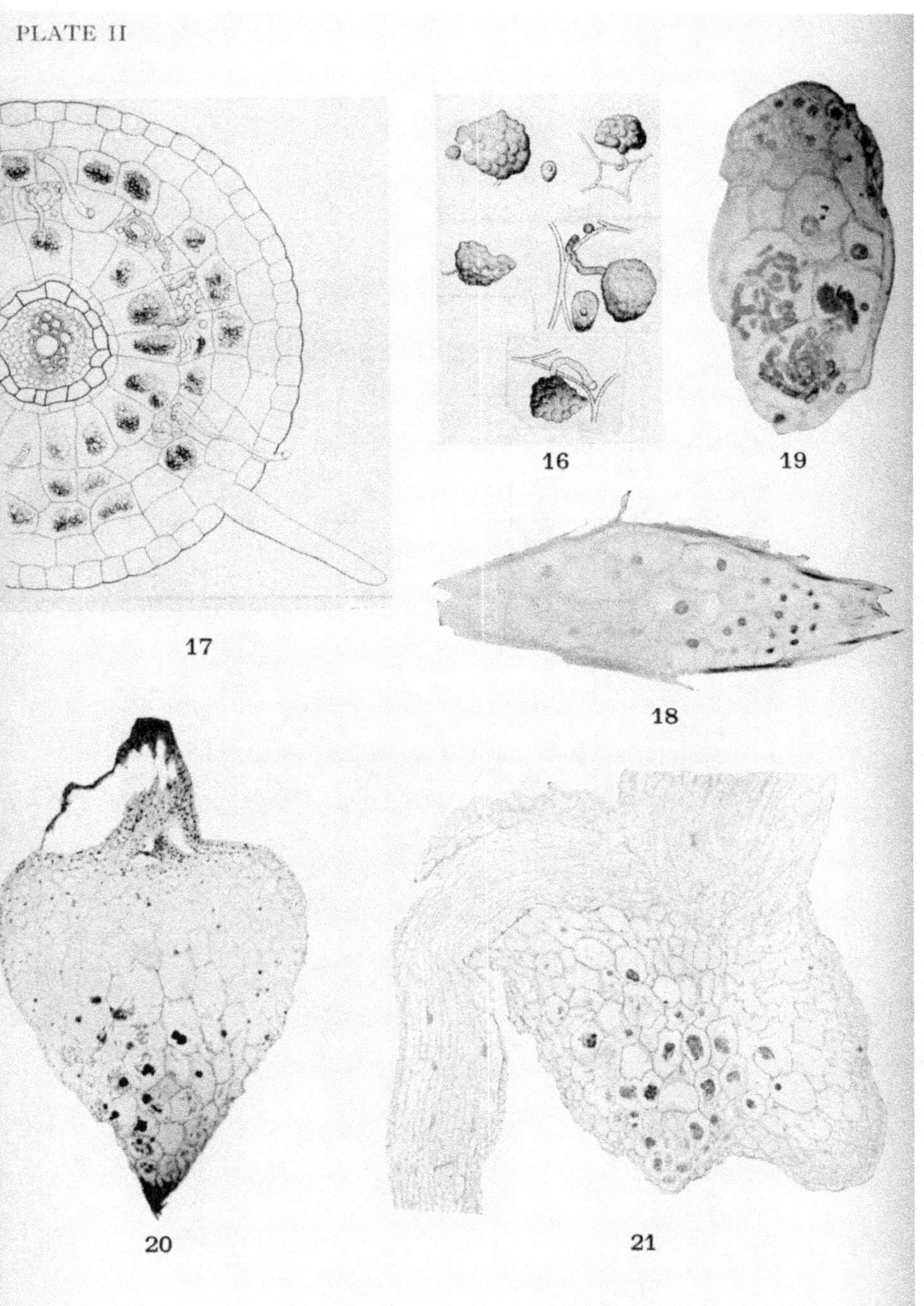

EXPLANATION OF PLATES II AND III

Plate II

Fig. 16. "Sporangioles" of different plants as figured by Janse. (From Janse, *Ann. Jardin Buitenzorg*. **14**, 1896–97.)

Fig. 17. *Polygala amara*. Transverse section of root with mycorrhiza. (From Marcuse, 1902.)

Fig. 18. *Odontoglossum* sp.: longitudinal section of seed. (From Ramsbottom, *Brit. Mycol. Soc. Trans.* 1922.)

Fig. 19. *Odontoglossum* sp.: longitudinal section of seed, nine days from sowing, showing infection of the larger cells at the suspensor end and the formation of "pelotons." (From Ramsbottom, *Brit. Mycol. Soc. Trans.* 1922.)

Fig. 20. *Odontoglossum* sp.: section of protocorms. The stem apex and first two leaves can be seen at the upper end. In many of the infected cells, the mycelium is already digested. (From Ramsbottom, *Brit. Mycol. Soc. Journal*, 1922.)

Fig. 21. *Odontoglossum* sp.: longitudinal section of young seedling after formation of first root. The root is uninfected; the dark patches in the root-cells are raphides. (From Ramsbottom, *Brit. Mycol. Soc. Journal*, 1922.)

Plate III

Fig. 22. *Platanthera chlorantha.* Tangential section of root showing "Pilzwirthzellen" and cells containing raphides. (From Burgeff, 1909.)

Fig. 23. *Platanthera chlorantha.* Radial section of root showing "Verdauungszellen." (From Burgeff, 1909.)

Fig. 24. Asymbiotic germination of Orchids. Seedlings of *Odontoglossum* sp. in organic nutrient 18 weeks from sowing. About natural size.

Fig. 25. Symbiotic germination of Orchids in horticultural practice. Seedlings of *Odontoglossum* sp. in sterilized compost infected by the endophyte, four months from sowing. × ½.

Fig. 26. Asymbiotic germination of Orchids. Seedlings of *Cattleya* and *Laelio-Cattleya* two years old on full nutrient without sugar. (From Knudson, *Bot. Gaz.* 1924.)

Fig. 27. *Gastrodia elata*: (1) Adult tuber with flower shoot.

(2) Inflorescence axis.

(3) Rhizomorphs of *Armillaria mellea* (*Rhizomorpha subterranea*) from soil about tubers of *Gastrodia.*

(4) Living stock of *Quercus serrata* with rhizomorphs of *Armillaria mellea.*

(5) Fruit bodies of *Armillaria mellea* from dead stump of Oak in neighbourhood of *Gastrodia.*

Fig. 28. (6, 7) Development of young tubers without association with fungal symbiont: (*a*) at end of May; (*b*) at beginning of August; (*c*) at beginning of April in the following year.

(8) Rhizome-like tuber of *Gastrodia.*

(9–12) Small tubers making organic connection with rhizomorph strands (collected end of May).

(13) Adult flowering tuber occasionally attacked by rhizomorphs.

(14) Mycorrhiza formation in tuber laid under Oak tree in May, showing vigorous development and formation of offsets (observed in September).

(15) Potato tuber attacked parasitically by rhizomorph of *Armillaria mellea,* showing discoloration and collapse of affected tissues.

(16) Section through affected tissue of same showing distribution of strands as "*Rhizomorpha subcorticalis.*" (All from Kusano, *Journ. Coll. of Agric. Tokio,* 1911.)

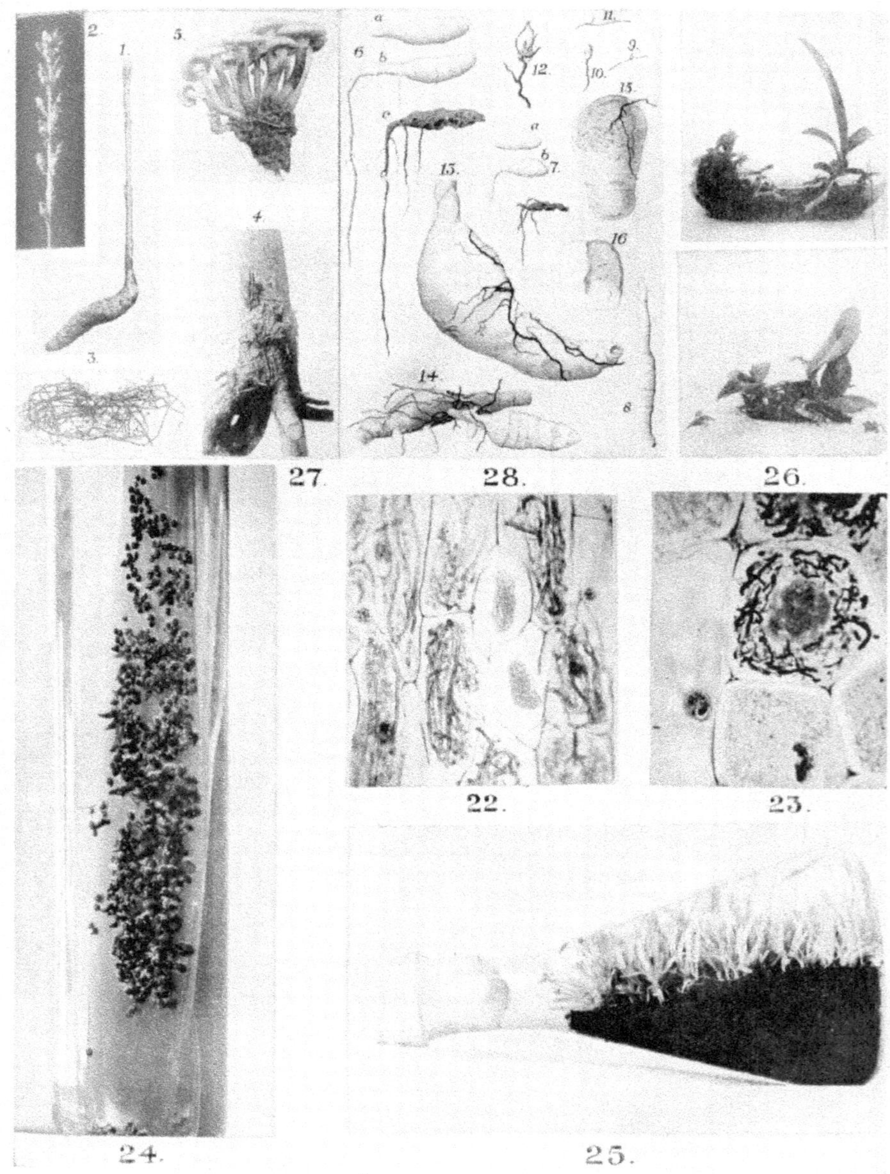
2.
1.
5.
4.
3.
a
6
b
c
12.
11.
9.
10.
15.
a
b
7.
13.
16
14.
8
27.
28.
26.
22.
23.
24.
25.

CHAPTER VI

ERICACEAE

Ericaceae—Ternetz—Rayner: seed infection and the obligate relation in *Calluna*; isolation of the endophyte, and its behaviour in pure culture; shoot infection; cuttings; the cytology of digestion in the mycorrhiza cells; nitrogen fixation; nutrition—Christoph—Other members of Ericaceae: *Erica*; *Pernettya*; *Vaccinium*; *Arbutus*; *Epigaea*—Pyrolaceae.

ALTHOUGH cited by Frank as likely to exhibit specialised relations with their root-fungi, the members of Ericaceae were relatively neglected by contemporary workers at mycorrhiza, a fact possibly not unconnected with the recognition of difficulties of technique due to fineness of the roots, abundance of oil in the tissues, and comparative difficulty of manipulation under experimental conditions. Isolated observations on the mycorrhiza of *Calluna*, *Vaccinium* and *Andromeda* were recorded by Frank (1892 *b*), who described and figured infected roots, and later by Magnus (1900) and others, but no experimental work was attempted. Even in respect to root structure, the published records were singularly incomplete, confined to a limited number of genera and deficient in reliable cytological details.

The first serious contribution to the literature of mycorrhiza in the group was made by Ternetz (1907). In an earlier paper (1904), this author had published an account of a fungus isolated from a peaty soil and believed to possess the power of fixing atmospheric nitrogen. The prevalence of ericaceous species upon such soils and their invariable development of endotrophic mycorrhiza, together with the observations just mentioned, led to an attempt to isolate the root-fungi of various members of the group with a view to testing their capacity in respect to nitrogen fixation. The absence of the nitrogen-fixing bacteria, *Azotobacter* and *Clostridium*, from heath and moorland soils put on record by Beijerinck (1901) and Christensen (1906) and confirmed by Ternetz, suggested the possibility that the activities of these bacteria were replaced, in soils of this type, by those of soil fungi that might or might not exhibit symbiotic relations with the roots of the vascular plants present.

From the young roots of five ericaceous species, *Oxycoccus palustris*, *Andromeda polifolia*, *Vaccinium vitis idaea*, *Erica tetralix* and *Erica carnea*, Ternetz isolated five pycnidia-forming fungi showing the general characteristics of the genus *Phoma*, to which were given the names *Phoma radicis oxycocci*, *P. r. andromedae*, *P. r. vaccinii*, *P. r. tetralicis* and *P. r. ericae*. Possessing general characters in common, these forms exhibited minor morphological and physiological differences when cultured; moreover, no two of them were ever isolated from a single culture, a fact regarded as significant evidence of their specific relations in view of the circumstance that roots of two or more of the ericaceous species used for their isolation had been growing in close proximity in the field. Although the specific names just quoted implied identity with the endophytes, no proof existed that this was the case, as indeed was freely admitted by Ternetz. The necessary evidence could be supplied only by inoculations from pure cultures into fungus-free seedlings raised from sterilised seeds, with subsequent formation of mycorrhiza. Unfortunately all attempts to provide such proof were unsuccessful; in some genera, e.g. *Vaccinium* and *Andromeda*, the sterilised seeds did not germinate; in others, e.g. *Calluna*, the seedlings derived from them showed after some weeks a typically infected condition of the root-cells. Seedlings were never obtained free from mycorrhizal infection, a circumstance which suggested to the author the possibility of seed-coat infection sufficiently extensive to resist the seed sterilisation methods employed.

All the fungi isolated by Ternetz were tested for nitrogen-fixation by cultivation in liquid media free from combined nitrogen, aerated by means of a slow stream of purified air passed through the culture flasks. After four weeks' growth, the nitrogen contents of mycelia and culture fluids were estimated separately by the Kjeldahl method. All the cultures yielded evidence of nitrogen-fixation, although in very different degree. The fungi extracted from *Oxycoccus*, *Vaccinium* and *Andromeda* showed the greatest activity, the values recorded after four weeks' growth being respectively 18 milligrams, 22 milligrams and 11 (10·92) milligrams of nitrogen assimilated for each gram of dextrose used.

As compared with the values obtained by Winogradsky (1902) for *Clostridium pasteurianum* and Gerlach (1902) for *Azotobacter chroococcum*, the absolute amounts of nitrogen fixed by these fungi are relatively insignificant. On the other hand, they worked far more economically than did the bacteria in relation to the amount of sugar consumed during growth, and in this respect were the most efficient nitrogen-fixers yet recorded. For purposes of comparison, Ternetz tested the possibility of nitrogen-fixation by *Aspergillus niger* and *Penicillium glaucum* and obtained evidence of a relatively feeble capacity to utilise gaseous nitrogen on the part of these species also, a result subsequently corroborated by Froelich (1907). The comparative values obtained for all the fungi named and also for *Clostridium* and *Azotobacter* are set out in the table now reproduced (Table I).

TABLE I.

(From Ternetz, *Jahr. f. wiss. Bot.* Bd. XLV, p. 388, 1907.)

Name des assimilierenden Organismus	Dauer des Versuches Tage	Gebotene Dextrose gm.	%	Verarbeit. Dextr. gm.	N-Gewinn mg.	Ass. N pro 1 gm. verarb. Dextr. mg.	Bemerkungen
Clostridium Pasteurianum (Winogradsky)	20	40	4	40	53·6	1·34	
Clostridium Pasteurianum (Winogradsky)	20	20	2	20	24·4	1·22	
Clostridium Americanum (Pringsheim)	30	1·25	0·25	1·25	4·6	3·7	
Clostridium Americanum (Pringsheim)	30	5	1	3·01	8·2	3·01	
Azotobacter chroococcum (Gerlach und Vogel)	35	5	0·5	5	42·7	8·56	Mittel aus 2 Besten
Azotobacter chroococcum (Gerlach und Vogel)	35	12	1·2	12	127·9	10·66	
Aspergillus niger (Ternetz)	28	7	7	1·1	1·9	1·71	
Penicillium glaucum (Ternetz)	28	7	7	0·7	2·8	3·8	
Phoma rad. Oxycocci (Ternetz)	28	7	7	0·85	15·3	18·08	
Phoma rad. Andromedae (Ternetz)	28	7	7	0·67	7·3	10·92	
Phoma rad. Vaccinii (Ternetz)	28	7	7	0·71	15·7	22·14	nicht ganz bakterienfrei
Phoma rad. Tetralicis (Ternetz)	28	7	7	1	4	3·99	
Phoma rad. Ericae (Ternetz)	28	7	7	1·1	2·3	2·17	

Comparative cultures of the different forms showed that the amount of nitrogen fixed was independent of increase in dry weight, the two forms of *Phoma* which showed the greatest increase in dry weight after four weeks' growth giving the smallest values for assimilation of nitrogen. Of the latter, the greater part was contained in the nutrient fluid in each culture, only a small proportion being found in the mycelium. It is significant also that in no case was a supply of combined nitrogen required for active growth of the cultures.

Due precautions were observed by Ternetz in respect to technique and the adequate provision of check and control cultures, and the estimations were made with great care. Some doubt was expressed by the author respecting the adequacy of the Kjeldahl method for dealing with very small quantities of nitrogen, but it was concluded on the basis of an independent series of tests, that the method was satisfactory if the values actually obtained were treated as approximate only. The absence of any grounds for adverse criticism of Ternetz' experiments is endorsed by Duggar (1916) in a critical review of work dealing with the vexed question of nitrogen-fixation by fungi.

The path of the experimental worker who attempts to prove the assimilation of atmospheric nitrogen by micro-organisms is a particularly thorny one. The values obtained are of necessity relatively minute, and the possibilities of experimental error relatively great. Moreover, in the case of soil organisms, the matter is of great practical importance and very rigid proof is rightly demanded before the claims made can be regarded as definitely established. Recognition of the value of Ternetz' work on nitrogen assimilation by the endophytes of Ericaceae has been unduly delayed owing largely to the author's failure to establish the identity of the fungi isolated. Subsequent work has provided ample confirmation of her conviction that she was working with the true endophytes, and rendered possible a belated tribute to the skilful and accurate technique by which their isolation from the roots was successfully accomplished. It has also furnished corroborative evidence that the root-fungi of ericaceous species can assimilate atmospheric nitrogen.

Exact information respecting the bionomics of mycorrhiza in Ericaceae dates from the publication of a paper on *Calluna vulgaris* (Rayner, 1915).

Experimental work was initiated by a study of the edaphic relations of this species in a selected area, and an attempt to evaluate the significance of the calcifuge habit (Rayner, 1911). Field observations indicated that the failure of the plant to spread by seed in the area investigated was apparently strictly correlated with a rise in the calcium carbonate content of the soil, with which was associated a shifting of the reaction in the alkaline direction. Experiments were devised to investigate and analyse the soil factors concerned in this limitation of growth with the view of throwing light on the significance of the calcifuge habit in the group as a whole. Pot cultures in calcareous and non-calcareous soils yielded evidence of abnormal growth in the former as evidenced by reduced germination capacity, retarded rate of germination, arrest of root growth and distorted development of the growing region, and arrest of shoot growth with reduced size and red coloration of the foliage leaves. Intimately associated with the abnormalities of growth noted in plants grown in calcareous soils was the presence of dense colonies of bacteria on the surface of the roots, especially around the tips, and a marked diminution of vigour in the growth of the mycorrhizal fungus normally present. It was found, moreover, that similar growth reactions could be induced in seedlings growing in a favourable soil by watering with filtered extracts of calcareous soil.

At this stage of the investigation it was not clear whether the bacteria present upon the roots in calcareous soils were pathogenic agents directly responsible for the abnormal behaviour of seedlings, or were to be regarded merely as indicators of soil conditions unfavourable to growth and, incidently, to the maintenance of the favourable relations maintained with the mycorrhizal fungus under favourable soil conditions. Subsequent work with soil extracts under "pure culture" conditions proved the correctness of the second view (Rayner, 1921).

Laboratory experiments pointed to root-infection as an important biological factor controlling growth, and directed attention to the study of its origin and significance as an initial step towards an understanding of the somewhat specialised soil relations shown by *Calluna* and allied species (Rayner, 1913). Before, therefore, any specific edaphic problem involving soil preference could be attacked directly—before, indeed, such a problem could be clearly formulated—it was essential to prepare the way by undertaking an independent investigation into the physiology of the relationship between the *Calluna* plant and its mycorrhizal fungus. The difficulties and problems faced by Frank and all subsequent workers upon the biology of mycorrhiza immediately presented themselves. At what stage of development and from what source does root-infection take place? Is it possible to cultivate seedlings free from fungus infection? Can the root-fungus be isolated; how does it behave in pure culture; and what are its systematic affinities?

In the case of *Calluna* experimental research has provided answers to these questions and disclosed the existence of a remarkable and unsuspected biological relation between the green plant and its fungal associate, a relation resembling that found in Orchids in that it involves dependence of the seedling plant upon infection by an endophyte at a critical stage of development, but differing sharply therefrom in respect to the mode of infection and the widespread extension of mycelium throughout the shoot tissues (Rayner, 1913,

1915). As in Orchids, there is an obligate relation between the two organisms, and proper development of the seedling is bound up with infection by the appropriate fungus at germination. In Orchids, such infection is precarious, depending upon the presence of the root-fungus in roots of plants of the same species or in the soil around them in the proximity of germinating seeds; in *Calluna* infection is ensured by the presence of mycelium of the appropriate fungus upon the testa of the ripe seed. In Orchids, the distribution of the endophyte within the plant body is strictly limited; in *Calluna*, it is almost co-extensive with the tissues of the plant; mycelium extends throughout the root and shoot, spreads into the floral organs and passes from the columella and walls of the fruit to the testa of the developing seed. Only the resting embryo and endosperm are free from fungal infection; at all other stages of development and throughout its active vegetative existence every *Calluna* plant yet examined has been found to be a dual organism.

The full story of the association with the mycorrhizal fungus was pieced together as a result of experiments extending over a number of years. For a complete account of the researches, reference must be made to the papers cited and to those published subsequently (Rayner, 1921, 1922, 1925). The main facts are included in the following summary: The young roots of *Calluna* are extremely fine, consisting only of an axial vascular strand surrounded by a single layer of large cortical cells. Under normal conditions of growth each of the latter encloses a dense branch system of mycelium, the whole root forming a characteristic endotrophic mycorrhiza. Each intracellular hyphal complex is continuous with hyphae upon the external surface of the roots, some fine and hyaline, others brownish and of large diameter. The intracellular hyphae are of relatively large and uniform diameter with abundant oily contents, the latter being absent from the surrounding cytoplasm (Pl. V, Figs. 35, 36). The mycelium of the endophyte has not been observed to penetrate within the vascular cylinder of the root.

Throughout the growing season the mycorrhiza cells exhibit active intracellular digestion of mycelium with disappearance of the resulting—and presumably soluble—products. The nearer to the apical meristem, the more rapidly is digestion initiated. Its onset is marked by the usual signs of cell activity—increase in size and chromatic content of the nuclei often accompanied by deformation, "clumping" of the mycelium about the nucleus, disappearance of the sharp outlines of individual hyphae, and the gradual conversion of the hyphal constituents from the region of the nucleus outwards, to a structureless mass possessing strong stainability. The last stages in the process are marked by shrinkage of the nuclei and disappearance of the stainable contents (Pl. V, Figs. 36, 37). This intracellular digestion of mycelium is a continuous process observable throughout the vegetative season from early spring to late autumn. The proportion of cells in the active mycelial condition or undergoing digestion at any given moment varies with the time of year, the age of the root, and possibly also with the season and other external factors. The phenomenon of digestion was overlooked by most of the earlier observers, although Frank (1891) referred to the infected cells in roots of *Ledum* and *Empetrum* as showing similar staining reactions to those of Orchids, and (1892 *b*) figured a cell from the mycorrhiza of *Empetrum* showing what was probably the "clump" stage of digestion. Magnus (1900), fresh from observations on the Orchids, stated that only "Pilzwirthzellen" were present in the mycorrhiza of *Calluna*. Since the formation of mycorrhiza in *Calluna* is an annual phenomenon, the ultimate fate of the constituent cells is of little physiological importance; in roots which persist for a second season, they are exfoliated by the formation of pericyclic cork.

Certain recent observations on the formation of mycorrhiza by *Calluna* have an important bearing on the nature of the biological relation between the two organisms concerned, and incidentally explain the inability to observe root-infection recorded by certain workers. During a seasonal study of the mycorrhiza of *Calluna*, it was observed that the early spring roots were characterised by rapid growth and much lighter infection than was the case later in the season. This relative immunity to infection is believed to depend on the interaction of two sets of causes: (*a*) the differential effect of low soil temperatures on the growth of root-cells and mycelium respectively; (*b*) the existence of internal factors regulating the reaction of the cortical cells to infection, related in turn with certain external conditions controlling the rate of growth.

Partial immunity to infection, with a similar imperfect or abortive formation of mycorrhiza, was observed subsequently in healthy plants subjected to certain experimental conditions; for example, cultivation in a favourable soil or soil extract sterilised by heat, in sterilised sand irrigated with rain water or a weak solution of salts, and in agar nutrient. The conclusions drawn from these observations were thus summarised: "*The development of the endophyte in the mycorrhiza cells of* Calluna *is markedly inhibited by certain conditions of the rooting medium, and roots exposed to such conditions may appear to be uninfected*" (Pl. V, Fig. 40). It may reasonably be inferred that the formation of active mycorrhiza in *Calluna* is a "reciprocal phenomenon," conditioned not only by the activity of the fungus, but by the reaction of the root-cells to invasion, and by factors in the rooting medium directly related with the nutrition of either partner, or both (Rayner, 1925). One other conclusion stated in the same paper may be quoted; namely, that "fungal infection and the stimulus to development associated with it on the one hand, and the formation of root mycorrhiza on the other hand, must be regarded as distinct phenomena." There is no experimental evidence that the formation of mycorrhiza is obligate, although under ordinary soil conditions it is the natural sequence to seedling infection and may be closely bound up with the soil relations of the roots.

The possibility or otherwise of replacing the stimulus to development ordinarily supplied by infection by the addition of a suitable organic substance to the rooting medium has not yet been fully explored in Ericaceae. In view of the asymbiotic germination of Orchid seeds when supplied with appropriate organic material, it seems reasonable to infer that similar methods might be used successfully in Ericaceae, and that seedlings thus raised free from infection might grow satisfactorily without mycorrhiza. On the other hand, further experimental data are required respecting the length of time during which plants of various species of Orchid can be maintained in healthy growth lacking infection; the recent work of Huber (1921), on *Liparis loeselii*, has yielded results at variance with those of other workers and is possibly significant in this connection (see p. 103). This matter will be discussed more fully in a later section dealing with the physiology of nutrition.

The identification of mycelium in the shoot tissues of *Calluna* is a matter of great difficulty and can be accomplished only with the aid of a delicate and specialised technique. In the young seedling immediately subsequent to infection it is relatively easy to trace the path of the invading hyphae and occasionally to observe an intracellular complex resembling those in the root-cells in the leaf mesophyll. Such intracellular mycelium is subject to eventual digestion by the host cell. In general, the hyphae which traverse the stem tissues are excessively fine and restricted in position to the middle lamellae of the cell-walls. Otherwise, they show no special distribution but range throughout all tissues, vascular and otherwise (Pl. V, Fig. 38). Only in air spaces in the leaves and elsewhere, and in the dead tissues of the bark, are hyphae of normal size and structure to be found. The extension of mycelium throughout the shoot was experimentally verified by rooting cuttings under controlled conditions (Rayner, 1925). It was found that shoot cuttings, rooted under aseptic conditions in sterilised sand, formed a normal root system with infection of the type characteristic for plants growing in a sterilised rooting medium. Moreover, the endophyte has been extracted from similar shoot cuttings. These experiments with cuttings are believed to constitute a final reply to the claims put forward by Christoph (1921) in respect to non-infection of *Calluna* (see p. 103). (Fig. 29 and Pl. V, Figs. 39, 40.)

In the organs of the flower mycelium is abundant, and much of it undoubtedly belongs to the endophyte. In the tissues of the ovary wall, as in the vegetative axis, the hyphae are extremely fine and can be identified with difficulty; thence they spread to the outer cell layer of the seed-coats, a phase of infection possibly related to the breakdown of these cells in the final stages of seed development and one not difficult to observe in fortunate sections (Rayner, 1925, Plate VI, Fig. 1 *a*). Ripe seeds removed from unopened fruits carry fine mycelium, sometimes profuse, more often extremely scanty and difficult to put in evidence (Pl. IV, Fig. 32). At germination this mycelium becomes active and infection of the seedling root by fine hyphae can readily be observed in seeds germinating on filter paper. A similar infection stage has been noted by the writer in seedlings of *Pernettya* which had germinated viviparously within the fruit chamber, so repeating the observation made by Ternetz (1907) on *Andromeda polifolia*.

Fig. 29. *Calluna vulgaris*: cutting rooted in sterile sand under controlled conditions. Removed and photographed 40 days after insertion in tube. *S*, stem.

By careful methods of sterilisation it is possible to kill the mycelium of the endophyte upon the testa as well as free the seeds from casual contamination by micro-organisms. Seeds, so treated, germinate normally under aseptic conditions, but the resulting seedlings exhibit more or less complete arrest of shoot development with an inhibition of root formation which is practically complete.

If inoculated at planting from a pure culture of the endophyte, they develop normally (Pl. IV, Figs. 30, 31).

The endophyte has been isolated from young roots, from shoots, and repeatedly from seeds removed aseptically from unopened fruits. It is a pycnidia-forming fungus with the characters described by Ternetz for *Phoma radicis*, and the name *P. r. callunae* given to the form isolated by that author has therefore been accepted. The necessary proof of identity which Ternetz failed to obtain has been provided by re-inoculation into seedlings free from infection, and the identity of the forms isolated by that worker with the endophytes of a number of ericaceous species may be regarded, therefore, as finally established.

A sporing colony of the endophyte in pure culture is shown in Plate V; for the characters of the mycelium and pycnidia reference may be made to the original description and figures (Rayner, 1915, Plate VI, Figs. 8, 9). In old cultures the mycelium produces terminal and intercalary swellings which have also been observed in association with roots in nature. It can be cultivated on many different artificial media over a considerable range of *p*H values, but is liable to show marked change of behaviour, e.g. in relation to pycnidia formation, under prolonged cultivation on artificial nutrients. Its physiological properties are easily altered by changes in the substratum; in particular the power of promoting development in the seedling is readily impaired both by long cultivation outside the plant and by the nature of the food supplies. In general, such change is in the direction of active parasitism. (Pl. V, Figs. 41, 42.)

The evidence available bearing upon the physiology of nutrition will be discussed in a later section of the present work.

In view of Ternetz' researches on nitrogen assimilation by a number of forms of *Phoma radicis*, the reaction of the *Calluna* plant to substrata free from combined nitrogen is of special significance. Researches on this subject (Rayner, 1922 a) showed that "pure culture" seedlings, inoculated with the endophyte at planting and grown in an agar medium lacking combined nitrogen, throve as well as did the control seedlings to which potassium nitrate was supplied at the rate of 0·5 g. per litre. The seedlings not supplied with nitrate were, on the average, healthier than the controls, of bright green colour and of equally vigorous growth. Kjeldahl estimations of samples of the agar medium used for these cultures yielded negative results, and the conclusions were confirmed by subsequent work using a silica jelly substratum instead of agar agar.

Indirect evidence of nitrogen-fixation by members of the genus *Phoma* has also been supplied by Duggar and Davis (1916) in the course of a critical experimental study of the evidence for nitrogen-fixation by fungi, these authors having adopted special precautions to avoid experimental methods open to criticism on the score of inaccuracy. Among the species investigated were *Penicillium* spp., *Aspergillus niger* and *Phoma betae*. With regard to the two first-named genera, Duggar confirmed the experimental conclusions of Ternetz and others that these fungi can utilise atmospheric nitrogen to a very slight extent. In the case of *Phoma betae*, the values obtained ranged from 3·022 to 7·752 mg. per 50 mg. of culture fluid in 25 days, a known amount of combined nitrogen being supplied. These values are outside the range of experimental error and are of special interest for comparison with those obtained by Ternetz for the forms of *Phoma radicis*.

Differences of structure and behaviour were noted by Ternetz in the forms of *Phoma* isolated from different ericaceous species. This evidence of specificity in the endophyte is supported by the observed impracticability of using the *Calluna* endophyte for promoting germination in other genera, for example, in *Erica tetralix*, *Erica cinerea* or *Pernettya mucronata*.

Mention must here be made of the discordant experimental results reported subsequently by Christoph (1921) regarding fungus infection in members of Ericaceae. In respect to *Calluna vulgaris* and *Erica carnea*, it has been put on record by this observer that seedlings both from sterilised and unsterilised seeds remained free from fungus infection when grown on thoroughly sterilised soil from a *Calluna* station, while similar seedlings grown on the same soil untreated rapidly developed mycorrhiza of the characteristic type.

From the time of Frank onwards, the members of Ericaceae have been regarded as typical examples of obligate mycotrophy. Christoph rejects this view and reports that in very dry places, plants of *Calluna vulgaris* and *Erica carnea* are often found lacking mycorrhizal infection altogether, while in pot cultures allowed to become dry the fungus soon disappears. In respect to this, it was not stated whether plants were kept under observation during the whole of the growing season. Moreover, by rooting cuttings in sterilised soil, he obtained plants that were believed to show complete freedom from root-infection. It was inferred by Christoph that infection took place only from the soil, thus precluding the possibility of seed-coat infection.

Unfortunately, no satisfactory evidence has been provided that the seeds used for these experiments had been effectively sterilised, nor has adequate proof been supplied that a fungus isolated from *Calluna* roots was the true endophyte. On the other hand, it is believed by the present writer that the supposed absence of infection from roots of seedlings and cuttings grown in sterilised substrata, and from plants occurring in nature in dry situations, is adequately explained by the partial suppression of mycorrhiza formation under these conditions. Infection takes place inevitably—in the case of seedlings, from the seed-coats, in the case of cuttings, from the stem tissues—but the normal mycorrhizal condition is not established under certain well-defined rooting conditions. The invariable presence of mycelium in the root-cells and elsewhere in such plants can be demonstrated by the aid of a careful technique (cf. Pl. V, Fig. 40).

It is clear that, in nature, ericaceous seedlings germinating in the neighbourhood of the parent plants will always be liable to infection from the soil in addition to that derived from mycelium upon the seed-coats. A detailed criticism of Christoph's work will be found in two papers by the present writer (Rayner, 1922 *b*, 1925).

Infection of the ovary tissues by mycelium has been recorded for a number of other ericaceous species of widely separated affinities (Rayner, 1915). It seems likely, therefore, that the habit described for *Calluna* is common in the family Ericaceae, although it may be manifested in different forms and the degree of dependence of seedling development upon infection may vary within wide limits. Experimental evidence has already been secured that in certain species, e.g. *Erica cinerea*, *Erica tetralix* and *Pernettya mucronata*, development of the seedling, lacking infection, progresses no further than in *Calluna*. On the other hand, it has been stated recently by Melin (1921) that the seeds of a race of *Azalea mollis* investigated by him are free from mycelium. This, if correct, should involve a capacity for independent development in the resulting seedlings.

For various reasons the genus *Vaccinium* is of rather special interest. The systematic affinities of the Vaccinioideae are somewhat uncertain, while the edible fruits produced by many species of *Vaccinium* constitute a claim to interest on the economic side; moreover, it was specifically stated by Stahl (1900) that roots of *Vaccinium myrtillus* remained entirely free from fungus infection when plants were raised from untreated seeds in sterilised soil, and he added to this record the following generalisation:—"Während manche obligaten Mycorhizenpflanzen, wie wir früher gesehen haben, der Anzucht aus Samen und der Kultur grosse Schwierigkeiten bereiten, lassen sich die Ericaceen auch ohne Gegenwart von Wurzelpilzen unschwer kultiviren, und ihre Samen gehen, zwar oft langsam, aber in grossem Procentsatz und sicher ohne Mitwirkung symbiotischer Pilze auf." Since the roots of *Vaccinium myrtillus* under normal soil conditions possess well-developed mycorrhiza of the usual ericoid type, this statement, if correct, precludes the possibility of infection from mycelium present on the seed-coat, and involves direct infection from the soil in the manner believed at one time to be invariable in mycorrhiza plants.

With the view of elucidating the exact condition in the genus, a series of experimental researches on species of *Vaccinium* have been carried out by the writer subsequent to the year 1915. The species used for the greater part of the work were *Vaccinium oxycoccus* (*Oxycoccus palustris*) and *Vaccinium macrocarpum*. The results of these researches will shortly be published. They may be anticipated here by two statements bearing directly on the claim put forward by Stahl in the passage just quoted. It has been found that: (1) seeds of *Vaccinium oxycoccus* and *Vaccinium macrocarpum* are infected by their respective endophytes while still contained in the fruit, and (2) *infection* of the seed-coat of these species of *Vaccinium* is more deep-seated than in *Calluna*. These observations render it difficult to accept Stahl's statement that plants of a closely allied species of *Vaccinium* were "völlig pilzfrei" after five months' growth in sterilised soil (see p. 55). The fact of root-infection in *Vaccinium* was probably overlooked by Stahl, as more recently by Christoph in *Calluna*, because the formation of mycorrhiza is partially inhibited in the roots of both species when growing in a sterilised medium. Under these conditions the demonstration of mycelium in the mycorrhiza cells of the roots demands a more careful technique than was bestowed upon it by either of these observers.

Before leaving the subject of *Vaccinium*, the work of Coville on the Blueberry, *Vaccinium corymbosum*, may be mentioned (Coville, 1910, 1916, 1921). In an account of this species, cultivated for the sake of its berries on wet peat soils in the United States, Coville described and figured the mycorrhiza cells of the root but did not carry out experimental researches on the subject of infection. On general grounds of habitat and the known shortage of available nitrogen compounds in peat bogs, he favoured the view that this species of *Vaccinium* can indirectly utilise gaseous nitrogen through the intervention of its mycorrhizal fungus. The positive results obtained by Ternetz for the endophytes of other species of *Vaccinium* were cited by Coville in support of this opinion.

Only one other member of the Ericaceae, viz. *Arbutus unedo*, the Strawberry Tree, has been the subject of experimental investigation in respect to mycorrhiza. Whereas, in the case of most members of the group, the external structure of the roots is ordinarily unaffected by the development of the endotrophic mycorrhiza, many of the lateral roots of *Arbutus* develop into small tubercles (Dufrénoy, 1917).

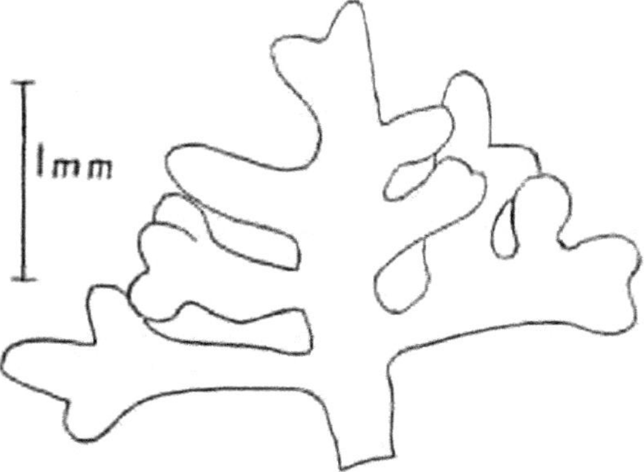

Fig. 43. *Arbutus unedo*: old branched tubercle. (From Rivett, *Annals of Botany*, 1924.)

Dufrénoy's material was obtained from plants growing wild in certain districts of south-western France, where *Arbutus* has a curiously limited distribution in the old Pine forests of Arcachon and La Teste but does not extend into the neighbouring Pine woods of Gascony.

The structure of the mycorrhiza in this plant has recently been re-investigated more fully by Rivett (1924), using material from seedlings raised in the British Isles. Rivett describes two distinct types of root in *Arbutus*, long roots which form the main root system, and root tubercles, shown by the author to be arrested laterals and sub-laterals of the season's growth (Figs. 43, 44). The long roots show a superficial network of mycelium with occasional penetration by hyphae; the type of infection bears some resemblance to the typical mycorrhiza of other members of Ericaceae but the intracellular development of the endophyte is relatively slight. The tubercles may be simple or branched, and are described as "relatively impermanent" organs. They originate from young lateral roots which suffer arrest of growth owing to profuse infection of the ectotrophic type. Subsequently, extensive endotrophic infection of the peripheral tissues occurs with rapid and complete digestion of the intracellular hyphae. Marked differences in the character of the mycelial growth, associated with the long roots and tubercles respectively, are described by the author who considers that the growth of the mycelium associated with roots is influenced by the presence of much mucilage on the emerging laterals, and by excretions from the ruptured tissues of the parent root. The presence of a mucilaginous sheath over the surface of the young roots is held by Rivett to be a normal feature of the root system in *Arbutus*, in disagreement with the view expressed earlier by Dufrénoy, who regarded it as a secondary feature associated with the presence of bacteria and algae. The mucilage is very variable in degree of development and may become conspicuous under dry rooting conditions. The roots of *Calluna* show a similar condition, the mucilage exhibiting precisely similar reactions to those described for *Arbutus*. In the former plant also it is abundantly developed in the neighbourhood of emerging laterals and around the young root-tips. In *Calluna*, as in *Arbutus*, it may become conspicuous when the rooting conditions are dry and may then serve to protect the young roots from drought. In both it is often well developed in perfectly clean young roots and doubtless plays a part in casual epiphytic infection by micro-organisms, perhaps also exerting a chemotactic influence on the initial infection of the seedling root by the mycelium of the endophyte in *Calluna*. (Pl. IV, Fig. 33.)

A curious feature of these tuberous mycorrhizas of *Arbutus* is the development by some of the hyphae in the outer part of the fungal sheath of stiff bristle-like setae up to 0·1 mm. in length. They occur only on young tubercles and on superficial examination resemble root-hairs (Fig. 45).

No information respecting the behaviour of uninfected seedlings of *Arbutus* is at present available. Ovarial infection has been recorded for the species (Rayner, 1915). Whether this, as in *Calluna*, carries with it a similar mode of infection at germination, and whether seedling development is bound up with such infection, are questions awaiting experimental solution. In view of the marked difference of

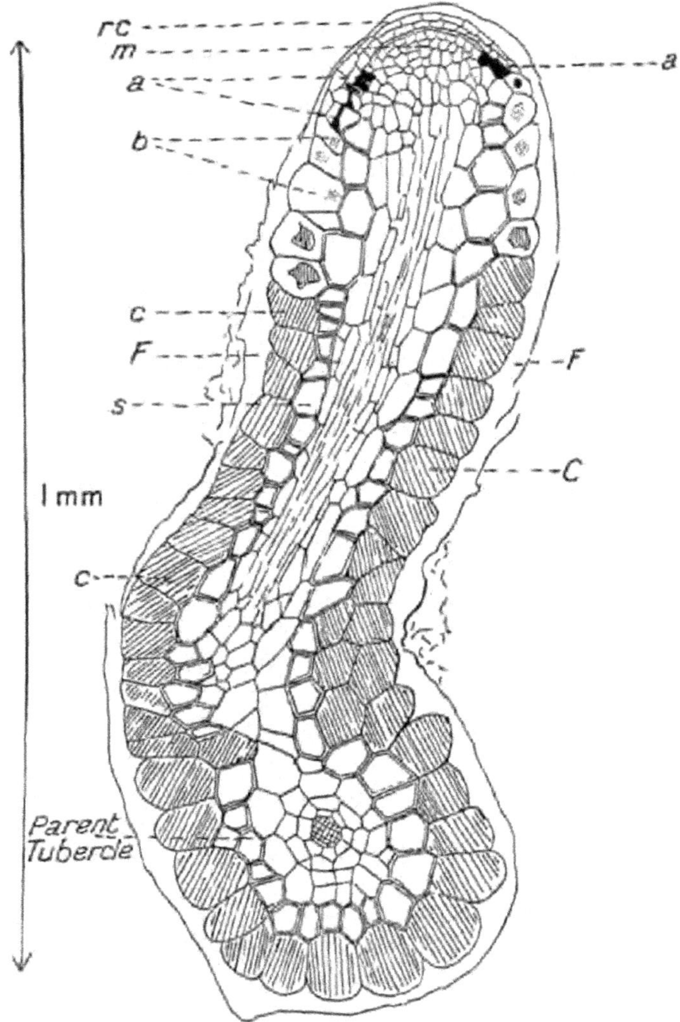

Fig. 44. *Arbutus unedo*: longitudinal section of young tubercle; *rc*, root-cap; *m*, meristem; *a*, first infected cells; *b*, cell with contents completely digested and large granular nucleus; *c*, reinfected cells; *F*, external hyphae; *s*, suberised layer. (From Rivett, *Annals of Botany*, 1924.)

structure and distribution of mycelium described it is tempting to speculate on the possible association of more than one fungus species with the roots of *Arbutus*. The presence of mycelium of more than one kind is not suggested by Rivett and must remain

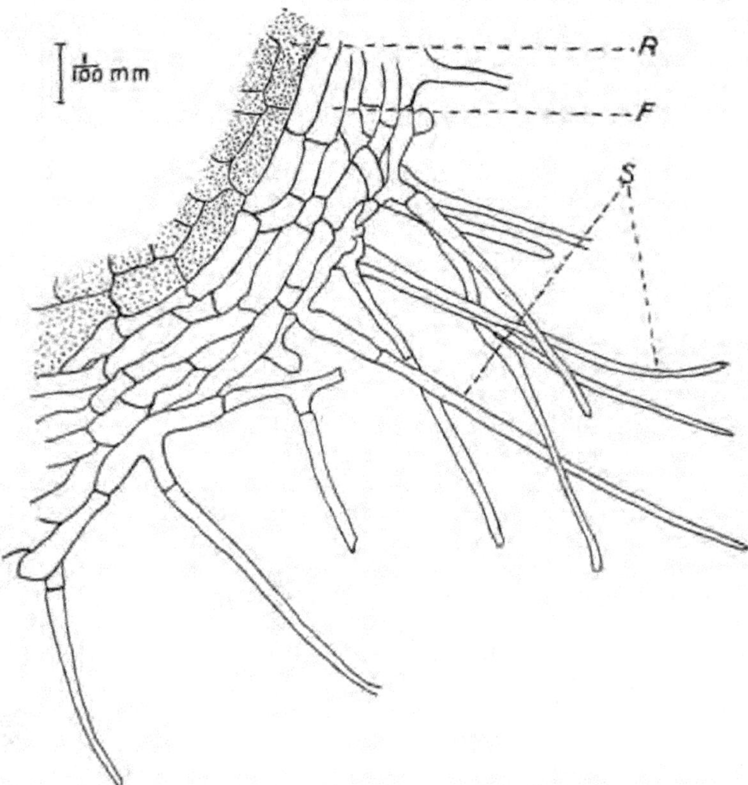

Fig. 45. *Arbutus unedo*: young tubercle with setae; *S*, setae; *F*, exterior fungal hyphae; *R*, root-cap cells of host. (From Rivett, *Annals of Botany*, 1924.)

a matter of speculation until the isolation and cultivation of the mycelium associated with the roots has been successfully accomplished.

Although *Arbutus* is exceptional in producing tubercles, the capacity for doing so exists in other species, e.g. in *Calluna vulgaris*.

The photograph in Pl. IV, Fig. 34 shows the formation of root-tubercles by a healthy plant of Ling from a typical callunetum in the south-east of England. The plant was vigorous and the roots showed no signs of pathogenic infection. The detailed structure of these tubercles has not yet been described.

Councilman (1923) has recently recorded the formation of mycorrhiza by *Epigaea repens* in America. Herbarium material from northern Japan showed an identical condition, a fact used by the author as an argument that the relation prevailed in the plants at their common source before the last glacial epoch. It was observed that the mycorrhiza cells sometimes contained material derived from alteration of the enclosed mycelium, and that such cells might undergo fresh infection. Presumably, therefore, intracellular digestion, of a similar nature to that described for *Calluna*, takes place regularly in *Epigaea*.

The mycorrhiza of *Epigaea* has not been the subject of experimental research, and Councilman has not at present correlated his observations with the results of experimental enquiry on *Calluna* and other members of Ericaceae.

For the smaller groups included in the Ericales, the available data respecting root infection and the biology of mycorrhiza where it occurs, relate almost exclusively to the Pyrolaceae. The members of this family are grouped in two sub-orders of which one, Pyroloideae, contains the chlorophyllous genera *Pyrola* and *Chimaphila*; the other, the Monotropoideae, includes *Monotropa* and the allied non-chlorophyllous genera usually described as holosaprophytes[1].

Mycorrhiza is formed by all members of the Pyrolaceae, and it has been pointed out by Henderson (1919) that it is possible to trace a series among the green forms showing a gradual transition from species with a relatively feeble production of mycorrhiza, e.g. *Chimaphila umbellata*, to others with heavier infection, e.g. *C. maculata*. In certain species of *Pyrola* some epidermal cells only are infected; in others, all the epidermal cells of the roots are filled with mycelium, there are indications of the formation of a fungal sheath about the root-tip, and rhizomes as well as roots are heavily infected. In the Monotropoideae, chlorophyll is not formed, and there is a

[1] The groups included by Engler in Ericales are as follows:—Clethraceae, Pyrolaceae, Lennoaceae, Ericaceae, Epacridaceae, Diapensiaceae. In other classifications the Vaccinioideae, a sub-order of Ericaceae, is treated as a separate family, Vacciniaceae, the genera *Pyrola* and *Chimaphila* are placed in Ericaceae, and the non-chlorophyllous members of Pyrolaceae (Monotropoideae) separated as a distinct family, Monotropaceae.

conspicuous fungal sheath about the roots showing differentiation into two zones.

In the paper to which reference has already been made, Christoph (1921) has given an account of his observations on members of the Pyrolaceae. Working with *Pyrola uniflora, P. secunda, P. minor* and *P. rotundifolia,* he reached the conclusion that, as in Ericaceae, no true symbiosis existed. Nor was he able to find any direct correlation between fungus infection and seedling development, or between infection and the formation of "coralloid" roots. Root-infection was described as involving an intercellular development of mycelium, but in *Pyrola* a typical ectotrophic structure was found only in those species which possessed rhizomes and only when growing in soils rich in humus.

The seeds of *Pyrola rotundifolia,* a species showing well-developed mycorrhiza in certain localities, germinated independently of infection, and the same was believed to be true for other species of *Pyrola* observed to germinate naturally on dry soil from which the fungus was judged to be absent. Seeds of the first-named species did not germinate upon sterilised soil, but success was achieved by the addition to the substratum of various organic substances, e.g. concentrated soil extract and peptone solution, the bestresults being obtained by supplying a mixture of these two substances. The effect of the root-fungus upon germination was apparently not tested.

It was believed that all species of *Pyrola* are liable to infection by closely related fungi, the mycelium of which showed "clamp connections." The mycelium associated with members of the Monotropoideae has no "clamp connections" and presumably, therefore, does not belong to members of the Basidiomycetes.

In respect to the biology of fungus infection in Ericales, Christoph concluded that, except in the Monotropoideae, there is no evidence of exchange of nutritive material; in all other cases, the endophyte is a harmless parasite.

In view of the predilection shown by many members of the Pyrolaceae for humus soils and their association with non-chlorophyllous species, the results recorded by Christoph are somewhat unexpected. Members of the family have so many features in common with Ericaceae and Orchidaceae, e.g. the possession of small seeds and undifferentiated embryos, and a marked susceptibility to fungus infection, that it would not have occasioned surprise to find a similar dependence of seedling development upon invasion by the root-fungus. Discussing the germination of "dicotylous saprophytes" in general, Goebel (1905) observed that it apparently took place only in very special surroundings:—"Probably the fungi which are found in the roots in symbiosis are essential."

The family Epacridaceae was one of those cited by Frank as likely to show specialised relations with root fungi. Alike in this group, in Diapensiaceae, and in Clethraceae, the mycorrhizal relations require investigation. So far as is known to the writer, the field of enquiry is still entirely unexplored in respect to the mycorrhiza of the plants belonging to these families.

EXPLANATION OF PLATES IV AND V

Plate IV

Calluna vulgaris

Fig. 30. Seedling raised from sterilised seed; growing in nutrient agar inoculated at planting from pure culture of the endophyte, 27 days from planting. *Inset*, seedling of same age uninfected.

Fig. 31. Seedling raised from sterilised seed; in aseptic culture on filter paper, 27 days after removal from seed-dish. *Inset*, seedling of same age inoculated at planting. *f*, mycelium of endophyte.

Fig. 32. Seeds removed from capsule before dehiscence, showing fungal infection of the seed-coats. *h*, hyphae.

Fig. 33. Tip of young root, showing mucilaginous sheath. *m*, mucilage.

Fig. 34. Base of main stem, showing formations of tubercle roots.

Plate V

Calluna vulgaris

Fig. 35. Mycorrhiza cells of root: mycelium in active condition; *m*, mycelila complex. × 1620.

Fig. 36. Mycorrhiza cell in very early stage of digestion; "clumping" of mycelium. ×1350. (From Rayner, *The British Journal of Experimental Biology*, 1925.)

Fig. 37. Mycorrhiza cells in advanced stage of digestion. ×900. (From Rayner, *The British Journal of Experimental Biology*, 1925.)

Fig. 38. Mycelium in stem tissues. Fine hyphae in cells of pith extending into wood. ×1600. (From Rayner, *The British Journal of Experimental Biology*, 1925.)

Fig. 39. Young root of cutting (cf. Fig. 29) rooted in sterilised sand, showing typical fungus infection. ×325. (From Rayner, *The British Journal of Experimental Biology*, 1925.)

Fig. 40. Cell from young root of cutting rooted in sterilised sand, showing intracellular mycelium and formation of "suppressed mycorrhiza." (From Rayner, *The British Journal of Experimental Biology*, 1925.)

Fig. 41. The endophyte of *Calluna*, *Phoma radicis callunae*: sporing culture in dextrose agar, three months old; *p*, pycnidia.

Fig. 42. *P. r. callunae*: mycelium from old culture (36 days) in dextrose agar.

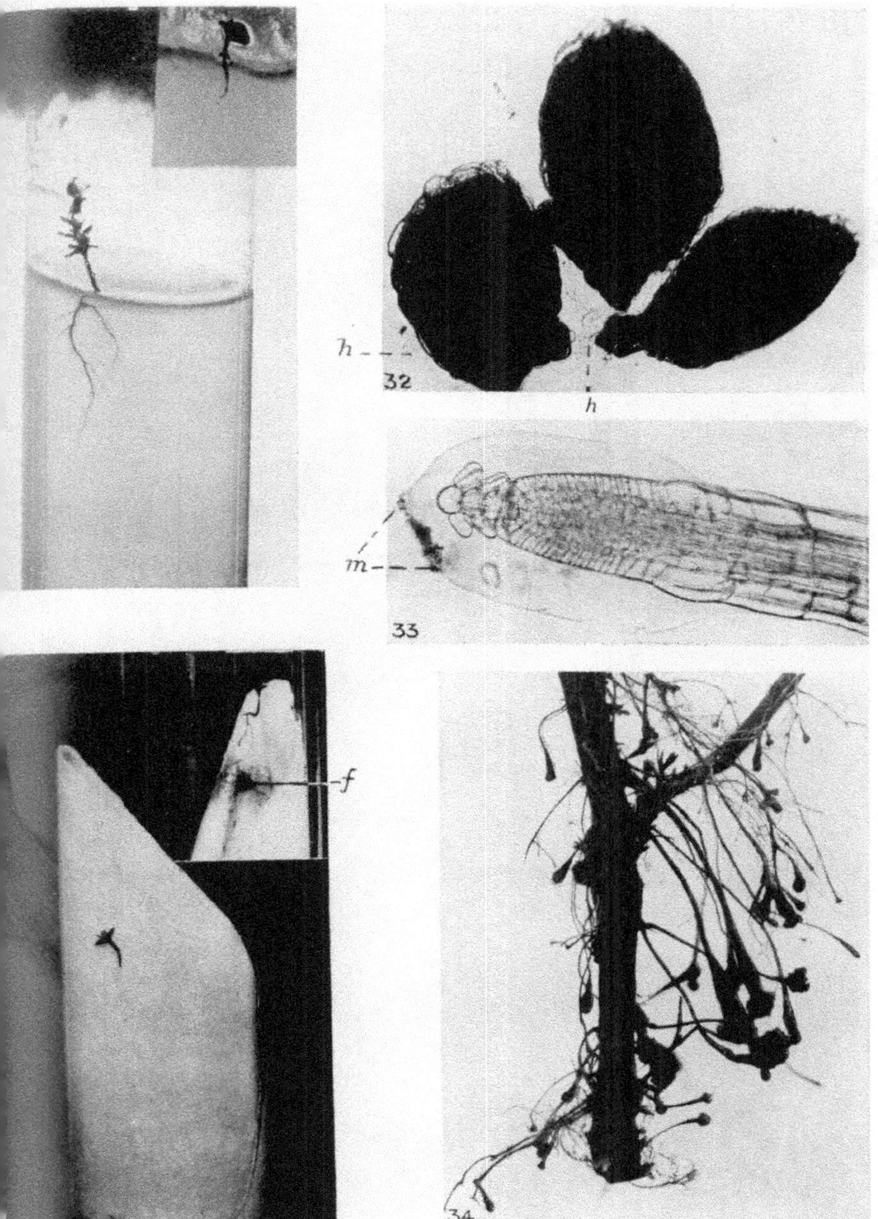
h
32
h
m
33
f
34

PLATE V

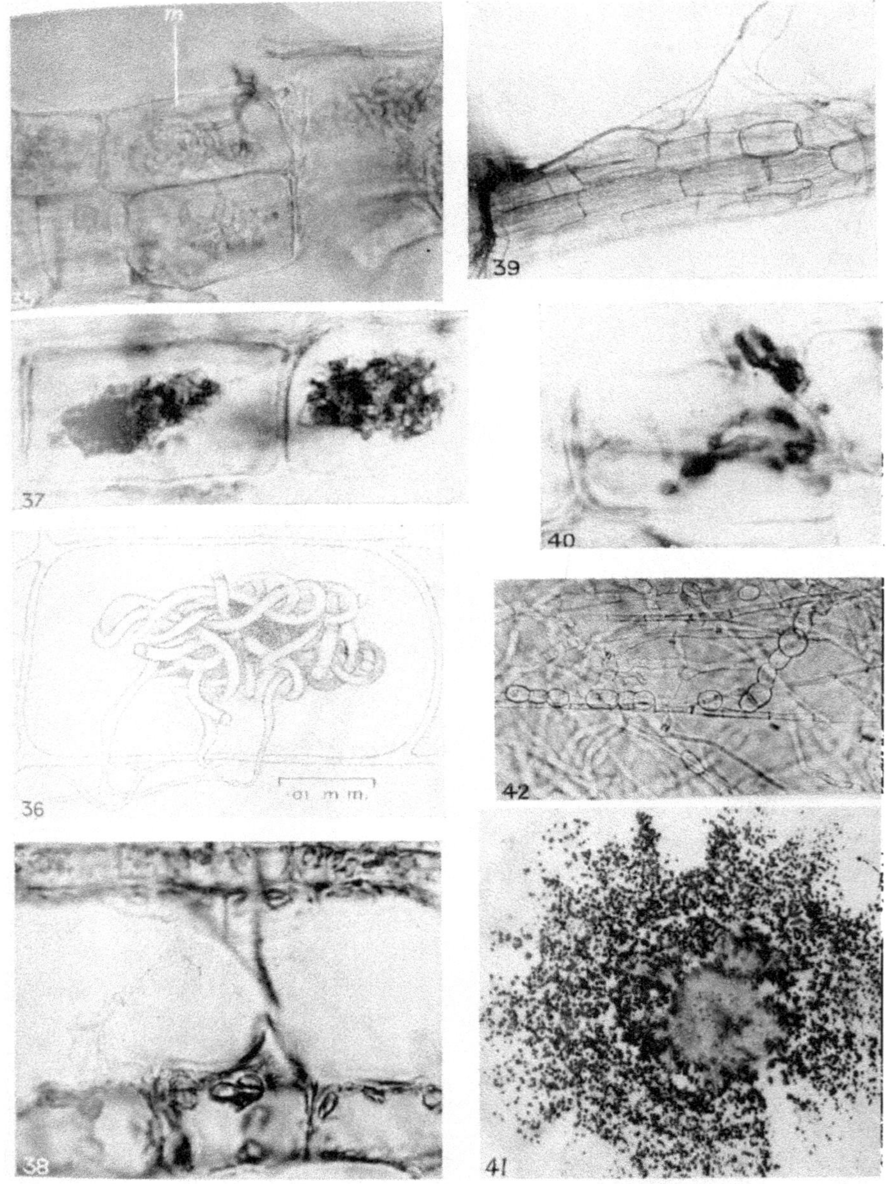

CHAPTER VII

RECENT RESEARCHES

Recent researches on endotrophic mycorrhiza other than that in Orchidaceae and Ericaceae—Peyronel's theory of "double infection"—Jones—Demeter—McLennan; fungus infection in *Lolium*—Constantin—Fossil mycorrhiza: Weiss; Osborn.

THE accounts of many independent observers will have made clear to the reader that the fungi concerned in the formation of endotrophic mycorrhiza fall into two fairly distinct groups in respect to their morphological characters. In one group are the Orchid endophytes, the mycelium of which possesses numerous cross walls and develops characteristic skeins or "pelotons" in the mycorrhiza cells. In pure culture these fungi form characteristic conidia and, whatever their immediate systematic affinities, they clearly belong to the Higher Fungi. In another group may be placed the endophytes of a majority of the herbaceous flowering plants, excluding Ericaceae and certain other obviously specialised groups. These are characterised by the formation of special organs—the "vésicules," "arbuscules," and "sporangioles" recorded and figured by many observers. Vesicles have been variously interpreted as reproductive organs or as vegetative structures functioning for the storage of reserves; arbuscules are generally regarded as of the nature of haustorial branch systems, recalling those formed by members of the Peronosporaceae; as a result of digestive activity on the part of the host cells they assume an appearance formerly interpreted as due to the existence of special organs—the "sporangioles" of Janse and other workers.

The widespread distribution of intracellular infection of the latter type was first put on record by Gallaud, who described and illustrated mycelium bearing arbuscules and vesicles from the roots of a large number of species of very varied affinities.

The presence of mycelium belonging to two distinct types in the mycorrhiza of flowering plants was suggested by the records of more than one observer and had led to some confusion in the literature, but the actual occurrence of such a condition was purely conjectural. Of observations pointing to such a composite type of infection may be mentioned those by Mollberg (1884) on *Epipactis* and *Platanthera* describing "vésicules" in the roots of those Orchids in addition to the characteristic intracellular "pelotons," and also those of Gallaud (1915) on the presence of "pelotons" resembling those of the Orchid fungi in the root cells of *Tamus* and *Psilotum*, side by side with typical arbuscules. Gallaud noted also that the roots of certain plants, e.g. *Allium sphaerocephalum* and *Ruscus aculeatus*, contained mycelium bearing arbuscules and vesicles with which were associated stromata similar to those produced by the Orchid fungi in pure culture. These and similar inconsistencies of structure were believed to depend upon the reaction of the endophyte to physiological differences in the root cells.

The observations recorded by Petri (1918–1919) on Vine and Olive were even more confusing. By placing mycorrhizas of these plants in moist chambers and allowing them to decay, this worker believed that he had secured evidence that the same mycelium which bore vesicles and arbuscules within the mycorrhiza cells, produced on the surface conidial fructifications quite analogous to those observed in cultures of the Orchid fungi by Bernard and Burgeff. Finding, as he believed, the same type of mycelium in plants so diverse as Vine, Olive and the Orchids, Petri deduced the conclusion that the fungi of endophytic mycorrhiza belonged to a single group, thus endorsing the views expressed earlier by Janse and Gallaud.

The theory involving explicit recognition of simultaneous infection by two distinct fungi is due to the Italian botanist Peyronel. The work of this observer on the fungi concerned in the formation of tree mycorrhizas will be discussed in a later section of the present work (Peyronel, 1920, 1921, 1922 *b*, 1924). In addition, he has recently reviewed and criticised the views expressed by other observers on the aspect of endotrophic infection now under consideration, and has carried out independent observations on the same subject. His own researches were designed to test the hypothesis of dual infection in endotrophic mycorrhizas showing, as he believed, two distinct types of mycelium. His observations have led to the publication of a new theory of endotrophic infection to which attention must now be directed (Peyronel, 1922 *c*, 1923, 1924).

From the time of his earlier researches on Wheat, Peyronel was struck by the dimorphic character of the mycelium in the mycorrhiza of that plant, the differences being sufficiently well marked, in his view, to suggest the presence of two independent fungus species. After long consideration he rejected this view, partly by reason of the arguments put forward by other workers, and partly because the two types of mycelium were not infrequently present in the same cells and did not lack structural features in common. His later observations, however, have led to a different interpretation of the facts, in sharp contrast with the opinions of Janse, Gallaud and Petri, and with his own earlier views. Peyronel's conclusions may be summarised briefly as follows.

The endophytes of Orchids on the one hand, and of the majority of flowering plants on the other hand, belong to two entirely different groups of the Fungi; the former are members of the Eumycetes, the latter show characters resembling those of the Phycomycetes. In each case, a number of biologic forms belonging to one or a small number of species are probably included. It is regarded as certain by Peyronel that endotrophic infection in a majority of mycorrhiza plants is of composite character, involving the presence of two endophytes; one, usually less developed, of the *Rhizoctonia* type familiar in Orchids, the other showing Phycomycete characters. The mycelium of the latter is recorded as widely distributed in damp humus soils, where it forms a fine network investing the roots of plants and passing from one to another. It flourishes under purely saprophytic conditions and may be found growing strongly in dead tissues *in situ* and in fragments of plant material in the soil. When living saprophytically it produces vesicles and lateral branch systems analogous to arbuscules.

In a majority of plants with endotrophic mycorrhiza, infection by the Orchid type of fungus supervenes upon that by the mycelium producing vesicles and arbuscules. The former develops chiefly in the outer cortex where it overgrows that producing arbuscules; it behaves rather as a *quasi*-parasite or saprophyte than as a true symbiont and is easily extracted and grown on artificial media.

In many endotrophic mycorrhizas, therefore, two distinct phases of infection may be recognised; an earlier, depending upon invasion by a fungus of the familiar arbuscule-producing type, and a later, in which a second organism of the *Rhizoctonia* type is also present. Fungi showing characters similar to the *Rhizoctoniae* of Orchids have been isolated by Peyronel from a number of different species of flowering plants, including several cereals, viz. *Triticum sativum, Zea mais, Hordeum vulgare, Solanum tuberosum, Nicotiana tabacum, Daucus carota, Beta vulgaris, Arum italicum, Euphorbia peplus, Circaea alpina* and *Saxifraga rotundifolia*. All the endophytes resembled one another and behaved in pure culture like those isolated from Orchids by Bernard, Burgeff, and Constantin and Dufond. Peyronel has accordingly accepted the generic name *Rhizoctonia* for these fungi.

The form isolated from Wheat has been re-inoculated into aseptic seedlings and has reproduced a mycorrhiza with typical "pelotons" but without the arbuscules and vesicles ordinarily also present in Wheat mycorrhiza (Fig. 46).

The endophytes belonging to the other group have not yet been isolated, but Peyronel has observed mycelium in dead roots and in soil similar to that in living roots. He has also identified it in cork and in fragments of bark of pot plants of *Citrus*, *Olea europaeus*, and *Morus nigra*. Moreover, when dead roots known to be infected were examined at intervals during the autumn and winter, the production of large numbers of vesicles could be observed. Peyronel's recent researches on dead roots of Wheat, Maize and wild grasses are

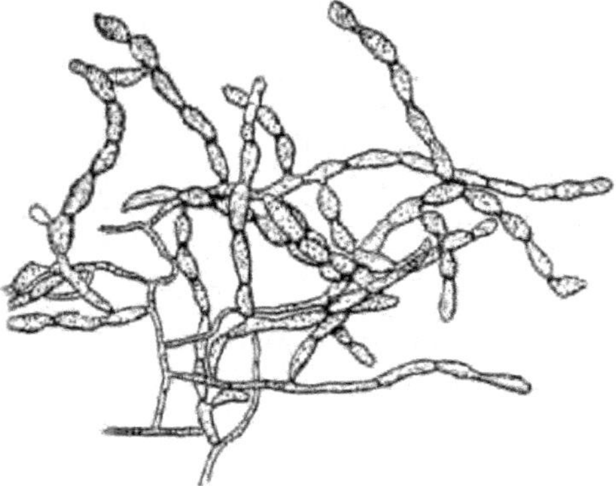

Fig. 46. Mycelium and moniloid chains produced in pure culture by a form of *Rhizoctonia* isolated from Wheat. Compare fig. 15 showing similar structure of an Orchid endophyte. Original × 375. Reduced about ½. (From Peyronel, 1924.)

reported to have established the direct relation of the vesicles with mycelium bearing arbuscules, so confirming his view that the former are sporangia of the same fungus in various stages of development. The spores eventually formed are constant in size and shape and are set free by changes in the sporangium wall (Fig. 47).

In the absence of sexual organs the evidence at present available is not regarded as sufficient to assign these fungal forms to a definite systematic position. It is suggested that they show affinities with the genus *Endogone*, a small group classed at one time with the Gasteromycetes or the Tuberaceae, more recently assigned to the Hemiascii by Baccarini (1903) and to the Oömycetes by Bucholtz (1911, '12). The sporangia of *Endogone macrocarpa* figured by Tulasne and reproduced by Schröter and Fischer are noted by Peyronel as remarkably like the vesicles produced by many mycorrhizal fungi.

Peyronel has likewise drawn attention to the isolation by Bernard (1911 a) of a fungus showing Phycomycete characters from the roots of *Solanum dulcamara*. A similar form was subsequently extracted from plants of the same species by Magrou (1921) who named it *Mucor solani*. He has pointed out also that the existence of what may be called a *normal radicicolous flora*—"flora radicicola normale"—on

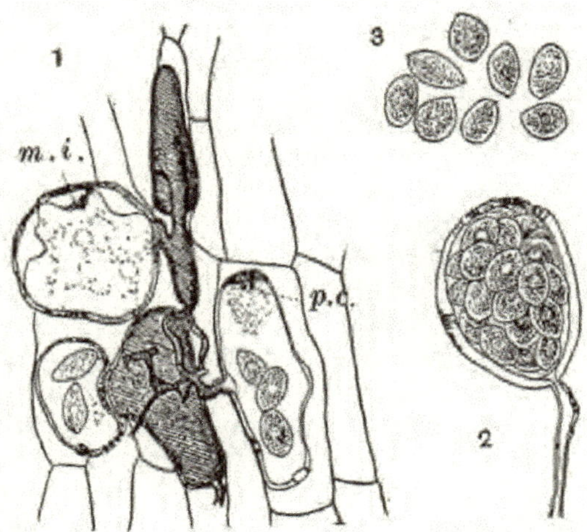

Fig. 47. Fructifications of an endophytic fungus bearing vesicles and arbuscules, from Maize and Wheat. 1. Group of ripe sporangia: *m.i.* internal membrane of a sporangium partly detached from the external membrane; *p.c.* conical processes on the internal surface of the sporangium wall. 2. Ripe sporangium with spores. 3. Spores. Original × 500. Reduced about ⅓. (From Peyronel, 1923.)

the roots of wild and cultivated plants at a certain stage of development is of practical importance in view of the fact that many of the fungi present are facultative parasites, the excessive development and virulence of certain constituents being possibly directly related to soil conditions. It is believed that the development of mycorrhiza in general is greatly accentuated in soils seldom cultivated and permeated with roots; hence its abundance in pasture fields and relative infrequency in cultivated plants and in ruderals.

It has recently been stated by Magrou (1921) that annual plants do not form mycorrhiza, or do so in a very transitory form. This statement is not acceptable to Peyronel, who has observed vigorous and persistent mycorrhizas in many annual plants under suitable conditions, e.g. in *Triticum aestivum*, *Hordeum vulgare*, *Zea mais* and *Secale cereale*. The theory of Bernard (1909a) associating the perennial habit in general with fungus infection has also been rejected as incompatible with his own observations.

Peyronel (1924) has recently published a long list of species in the mycorrhiza of which he had observed the dual type of endotrophic infection. They are distributed among 37 families, so that, assuming the correctness of these observations, this type of infection may be regarded as widespread among flowering plants.

It is of interest to compare Peyronel's observations on leguminous species in Italy with those recorded independently by Jones (1924) in the mycorrhiza of a number of cultivated legumes in the United States. Jones' observations were made first upon peas and subsequently extended to other crop plants. He noticed that a majority of apparently healthy lateral roots of the former plant showed conspicuous discoloration due to infection by a coarse non-septate mycelium with a profuse development of arbuscules and vesicles, i.e. of the characteristic Phycomycete type distinguished by Peyronel. Infection of this kind was observed also in greater or less degree in the roots of other leguminous crop plants, four of which, *Melilotus officinalis*, *Medicago sativa*, *Trifolium pratense* and *T. repens* are included in Peyronel's list (published subsequently) as showing dual infection. Neither in the account given by Jones nor in his figures is there any suggestion of the presence of another endophyte with distinct morphological characters. Since the *Rhizoctonia* type of infection follows the other, this omission may have been due to the age of the roots, or the distinction of the two types of mycelium may have been overlooked by the American observer. Attempts to extract the fungus from infected roots were unsuccessful.

The observations made by Jones are obviously incomplete, but are of interest as indicating the uniformity and widespread character of the Phycomycete type of mycorrhizal infection. While admitting that the relations of the root cells with the mycelium were not of the ordinary parasitic kind, he was not able to detect any beneficial effect upon the host plants, and indeed observed indications that leguminous crop plants benefited when protected from root infection by sterilisation of the soil before planting. Decisive evidence bearing on this matter can be obtained only by isolation of the endophyte or endophytes and "pure culture" experiments with infected and uninfected plants, and even then the results must be applied with great caution to plants growing naturally.

Under the name of "*Plasmoptysen-Mykorrhiza*" Demeter (1923) has recently described a type of root infection characteristic of Asclepiadaceae and Apocynaceae, with special reference to the mycorrhiza of *Vinca minor*. On experimental grounds he concluded that the mycotrophic habit is obligate in *Vinca* when growing under natural conditions, about two-thirds of the roots showing infestation. The manner of infection is related by the author to the presence of special cells, "Kurzzellen," in the sub-epidermal layer (exodermis) of the root, corresponding to the "cellules de passage" mentioned by Janse (1896–7) and the "Durchlasszellen" of Burgeff (1913) in other plants. Root infection is continuous throughout the year from early spring onwards. The general type of structure is that made familiar by Gallaud (1905): the mycelium, mainly intercellular in distribution at first, becomes intracellular owing to the development of haustorial branch systems or "arbuscules" which undergo degeneration later to structureless sporangioles. Terminal and intercalary vesicles of the usual kind are freely produced (see note 1 on p. 128).

The specific name given to this mycorrhiza by the author followed upon a cytological study of the arbuscules, of which the tips of the ultimate branchlets lose their identity, and become imbedded in a finely granular matrix staining deeply with haematoxylin. This structure is attributed by the author to bursting of the hyphal tips and liberation of the contents owing to the action of acid present in the cell sap. Alternatively, it is suggested that it may be brought about by an excessive rise in the osmotic content of the hyphae, due to enzyme activity and the intake of sugar (see also Burgeff, 1909). Demeter favours the former interpretation and has obtained evidence of similar behaviour on the part of the mycelium in artificial cultures when the fungus is grown on media containing hydrochloric acid at a concentration of about 0.025 *N*. The structures described were named "Plasmoptyse" by the author, who has accordingly adopted the name "Plasmoptysen-Mykorrhiza" for the type of mycorrhiza found in *Vinca minor*. Careful study of the figures supplied leaves some doubt as to how "Plasmoptyse" differ from the earliest phase in intracellular digestion, i.e. the initial stage in the conversion of arbuscules into sporangioles. In respect to the identity of the endophyte, the conclusions reached also provoke criticism. Fungal forms showing identical characters were isolated independently from plants growing in three separate localities. Under cultivation, this fungus, believed by Demeter to be the endophyte, showed morphological characters resembling those of the Orchid fungi and was named *Rhizoctonia apocynacearum* accordingly. On the other hand, it is recorded that marked dimorphism of the hyphae and great variability were exhibited in single cultures, according to the nature of the nutrient supplied. The author figures mycelium and conidia resembling those of the Orchid endophytes from artificial cultures and alludes also to the production of "arbuscules" and "Plasmoptyse" in similar colonies, "Diese 'Plasmoptyse' wurde an den in Reinkultur gezogenen Endophyten kunstlich ausgeführt und erfolgte im Optimum bei einer Säurekonzentration von etwa 0.025 *N* HCl..." but he does *not* figure mycelium bearing these structures and also conidia simultaneously. In view of Peyronel's (1924) recent conclusions respecting dual infection, i.e. the existence in many plants of secondary infection by a fungus of the *Rhizoctonia* type superimposed upon primary infection by an endophyte bearing vesicles and arbuscules, Demeter's views challenge attention. The production of conidia of the *Rhizoctonia* type in pure culture by a mycelium which also produces vesicles and arbuscules has never before been recorded and is in direct disagreement with Peyronel's conclusions. Indeed, the isolation and cultivation of an endophyte producing arbuscules is recorded for the first time in this paper. Inoculation experiments did not reproduce the mycorrhizal condition observed in roots growing naturally.

Peyronel (1924) has since included *Vinca minor* in his list of plants showing the dual type of infection. In view of this, and of the unsatisfactory character of the evidence supplied by Demeter in respect to the identity of the endophyte isolated by him, the conclusions reached by the latter require confirmation.

The claim put forward by Peyronel that "double infection" is a frequent and regular phenomenon in endotrophic mycorrhiza is of considerable interest as providing a reasonable explanation of certain observed inconsistencies of structure. It is supported by evidence of a suggestive kind but can hardly be regarded as definitely established. In the more specialised groups, evidence for its occurrence is at present almost negligible; records are extremely rare for Orchids and do not exist for members of Ericaceae. It may be that greater specialisation to the symbiotic relation carries with it an increased resistance to invasion by the "Phycomycete type" of mycelium. On the other hand, the fact that the latter has hitherto resisted isolation, except in the not very satisfactory case reported by Demeter, in itself constitutes evidence of relatively great specialisation in respect to the symbiotic habit, and is somewhat inconsistent with the wide saprophytic distribution reported in dead roots and organic soils. The endophytes of the *Rhizoctonia* group are described as showing morphological characters identical with those of the Orchid fungi. The latter show a high degree of specialisation in relation to particular hosts, although they grow as saprophytes with comparative readiness in pure culture. Peyronel himself has alluded to the *Rhizoctonia* forms from "double mycorrhizas" as behaving rather as saprophytes than as true symbionts, and he isolated them with comparative ease from a number of mycorrhizas. It is possible that edaphic factors may be of importance and it is evident that the whole subject offers a profitable field for further investigation.

The endophytic fungus of *Lolium* described by McLennan (1920) merits notice, more especially in view of the occurrence of seed-coat infection in Ericaceae. The existence of fungal infection in the grains of Darnel Grass (*Lolium temulentum*) had long been known. Papers dealing with the subject were published by Vogl (1898), Guérin (1898), Hanausek (1898) and Nestler (1898). Hiltner (1899), working at the physiology of the relation, believed that a case for nitrogen fixation had been experimentally established. Subsequently papers appeared by Micheletti (1901) and Freeman (1903, '04, '06), the latter recording sporadic infection of a similar kind in grains of other species of *Lolium*.

The whole subject has recently been reinvestigated by McLennan (1920), who has shown that the occurrence of fungal infection in the genus *Lolium* is wider and more constant than was believed. Infection of the grains of *Lolium temulentum* and *Lolium perenne* was found to be the rule, from whatever part of the world seed was obtained. The ultimate origin of infection in the genus is still unknown. Mycelium is present in the embryo-sac at, or immediately after, fertilisation, before the ovum divides. Infection spreads into the nucellus and carpel walls, but is limited in extent by the absorption of mycelium, which is apparently utilised as a source of food supply by the developing embryo. Ordinarily, a persistent layer is formed around the endosperm, although—"if the fungus does not keep pace with the absorbing power of the endosperm, no hyphal layer is formed in the ripe grain, but hyphae can then be found in the scutellum and embryo." After germination the mycelium grows *pari passu* with the seedling tissues, following the development of the stem apex.

The distribution of the mycelium is mainly intracellular and individual parenchyma cells of the young grass stems may contain a dense network of hyphae. At the flowering period mycelium is abundant at the base of the carpels, whence hyphae extend to the developing ovules. All attempts to isolate the fungus have been unsuccessful and no experimental evidence was obtained by McLennan in support of the view that nitrogen-fixation is a function of the endophyte. Nevertheless, it is believed that—"the association of the fungus with *Lolium temulentum* and *Lolium perenne* is probably a well-marked case of symbiosis, comparable in many respects with that met with in *Calluna vulgaris*." The formation of (root) mycorrhiza in *Lolium* had not been recorded by any observer at the time this paper was written. McLennan (1926) has since made a further contribution to the subject, recording the presence of endotrophic mycorrhiza in the roots of *Lolium* and describing in detail the cytology of root infection in one species, *L. temulentum*.

For some unexplained reason the formation of mycorrhiza by members of Graminaceae was long overlooked. Schlicht (1899) noted it in *Holcus mollis* and *Festuca ovina*: Schröter (1908) recorded his own observations and those of Schnellenberg on various alpine grasses. Jefferies (1916), describing the piliferous layer of the root of the Purple Heath Grass (*Molinia coerulea*), noted the frequent presence of mycorrhiza—"the mycelium being visible in the cells and on their surface, and occasionally penetrating to the cells of the second row." More recently, Peyronel (1922 *c*) described the mycorrhiza of Wheat and a number of other cereals as due to infection by a fungus producing the characteristic arbuscules and vesicles associated with endotrophic infection. To these records is now added that of McLennan for four species of *Lolium*, namely, *L. temulentum*, *L. perenne*, *L. multiflorus*, and *L. subulatum*.

The mycorrhizal fungus of Darnel Grass is of the Phycomycete type distinguished by Peyronel, characterised by the production of vesicles and arbuscules and the conversion of the latter organs to sporangioles as a consequence of digestive activity. The distribution of mycelium is similar to that recorded for *Arum maculatum* by Gallaud, the hyphae, intracellular in the outer layer of the root, becoming both inter- and intra-cellular in the deeper tissues. The mycorrhiza is described as "a particularly favourable one for studying the cytological details," and McLennan has drawn certain inferences respecting exchange of nutritive material from an intensive study of the infected cells. These will be considered later when reviewing the evidence relating to nutrition in mycorrhiza-plants in general. (Pl. VII, Figs. 61, 62.)

There is at present no evidence whatever bearing on the identity or otherwise of the two endophytes of *Lolium*, i.e. the mycorrhizal fungus proper and that associated with the fruit, seed and vegetative tissues of the shoot. The distribution of mycelium in the tissues of the seedling inevitably challenges comparison with that of various members of the Ustilaginaceae. It is greatly to be desired that future work may render it possible to isolate and cultivate the fungus or the fungi present in *Lolium*, and thus throw light on the systematic affinities of the endophytes and provide further evidence as to the biologic relations existing in this interesting case of symbiosis.

Another remarkable case of endophytic fungus infection other than that responsible for mycorrhiza formation has been reported in the family Bromeliaceae. On the older view, the epiphytic members of this group were reported to obtain the requisite nutritive materials from the decomposition products of vegetable remains and small insects caught in the gummy secretion of the leaves. Certain recent observations on *Tillandsia dianthoides* provide evidence that this, at least, is not the only explanation. A plant of this species, hanging freely from a wire, was kept under observation for fifteen years and observed to grow and flower regularly; no evidence was found that any mechanism existed for holding water or for capturing and digesting insects. Like other epiphytic bromeliads, this species of *Tillandsia* forms scale-like structures or "lepidotes" upon the surface of the shoot. On investigation, these proved to be small receptacles connected with deep-lying groups of cells. The cavities, in addition to the presence of organic and inorganic detritus, showed regular infection by a septate mycelium present in relatively great abundance. The fungus responsible was isolated and identified as a species of *Volutella*. It is assumed that the relation between the *Tillandsia* plant and the fungus is one of beneficial symbiosis, the latter supplying nitrogenous food material to the green plant (Dubois, 1925 *a*, *b*, *c*).

Chiovenda (1920) has recorded the rare fungus *Myriostoma califorme* for the second time in Italy. It was found closely associated with the roots of *Polygonum persicaria*, in which it was believed to be responsible for the formation of endotrophic mycorrhiza.

Ramsbottom (1922) has drawn attention to the association of small undifferentiated seeds with obligate mycotrophy in the families Orchidaceae, Burmanniaceae, Ericaceae, Pyrolaceae, and Gentianaceae, and has noted the numerous records of difficult or unsatisfactory germination and their possible significance in relation to fungus infection. The mycorrhizal associations existing in Gentianaceae have been insufficiently explored and would probably repay experimental study. Many of the members of this group are plants showing marked edaphic peculiarities, and not a few of them are difficult subjects in horticultural practice. Frank (1887*b*) and Schlicht (1889) cited *Menyanthes trifoliata* as a typical case of non-infection, and the latter even mentions Gentianaceae as a group in which he had been unable to find infection in the plants selected for examination. "Von den Familien aus denen ich Pflanzen untersucht habe, konnte ich eine Verpilzung der Wurzeln nicht beobachten bei folgenden: Crassulaceae, Scleranthaceae...Gentianeae...Cariceae."[1] Stahl (1900) alluded to the existence within the group of species recorded as free from infection, e.g. *Limnanthemum* and *Menyanthes*; together with the holosaprophytic species of *Voyria* described by Johow (see Pl. I). Stahl himself provided a long list of species—alpine and other species of Gentian, *Erythraea centaurium* and *Chlora perfoliata*, in which he had observed regular infection. Schröter (1908) afterwards confirmed these observations for a number of alpine Gentians. It was suggested by Holm (1897) that the curious gentianaceous species *Oblaria virginica* was a connecting link between the typical green members of the family and the saprophytic forms *Voyria* and *Voyriella*. In *Oblaria* the green colour of the shoot is masked by the presence of anthrocyanin in the epidermis; the root system shows a "coralloid" habit and typical endotrophic infection; there are no root hairs.

The relation of mycorrhiza to pathological conditions in roots has interested mycologists from the time of Theodor Hartig onwards. Discussing this matter in reference to certain plant diseases, especially those of Sugar Cane, it has been pointed out recently that disturbance of the normal activity of the mycorrhizal fungus may lead to secondary invasion of roots by bacteria or parasitic fungi, e.g. species of *Marasmius* or *Rhizoctonia* (Constantin, 1924). Parasitic members of the latter genus cause much damage to plants and some of the pathogenic forms closely resemble the endophytes of Orchids (Constantin, 1925). Observations of this kind raise questions of great practical interest, especially in relation to tropical agriculture. It may be possible to correlate them with isolated observations by other workers, e.g. those on the bacterial investment of ericaceous roots in calcareous soil (Rayner, 1913), and also with the formation of "Pseudomykorrhiza" in the Conifers, and the invasion of roots of

[1] The only member of Gentianaceae mentioned in Schlicht's list of species examined is *Menyanthes trifoliata* !

these trees by fungi of a more or less pathogenic character under certain soil conditions (Melin, 1923 *a*, 1925 *c*).

In view of the possible affinities of the family Aristolochiaceae with parasitic or semi-parasitic groups like the Rafflesiaceae, Loranthaceae and Santalaceae, the formation of mycorrhiza by *Asarum europaeum*, recorded by Schwartz (1912), is noteworthy.

In this plant, an uncommon and somewhat doubtful member of the British flora, the young adventitious roots are reported to form endotrophic mycorrhiza of a typical kind. It is apparently inconstant in appearance, being absent from individual plants from certain stations and also from *Aristolochia sipho*, another member of the same group. Schwartz described and figured intracellular infection of the inner layers of the cortex, with disappearance of starch from infected cells. It was noted that the endophyte resembled those of *Thismia* and *Neottia*. In the outer cells of the infected zone the hyphae formed rather loose coils or "pelotons" while the cells of the innermost cortical layer showed typical intracellular digestion of the mycelium. The formation of "arbuscules" was not observed, but the author recorded the presence of spherical and pear-shaped vesicles of the familiar type. In the light of Peyronel's "double infection" hypothesis and in view of Schwartz's tentative suggestion that in *Asarum* the mycelium bearing these may belong to a different fungus, the mycorrhizal condition in *Asarum* might repay further investigation[1].

The part played in nutrition by the fungi present in endotrophic mycorrhiza, and the biological significance of root infection in relation to differences of soil and other external factors continues to be a subject of controversy. There is a strong *prima facie* case for a beneficent action upon the nutrition of the host based upon cytological and microchemical evidence. To supplement this, data derived from pure culture experimental researches on certain groups is now slowly accumulating. The whole subject will be dealt with more fully in a later chapter.

Fossil Mycorrhiza

In view of the wide distribution of mycorrhiza at the present day and also of its biological significance, observations suggesting its occurrence in fossil plants are of considerable interest. The evidence relating to fossil fungi in general was summarised by Seward (1898), who concluded that there is fairly good evidence for the existence of phycomycetous fungi in Permo-carboniferous times, the earliest

[1] See note 2 on p. 128.

reliable records for the Higher Fungi occurring much later, in Post-paleozoic or even Tertiary times.

The "coal-balls" from which so much of the evidence relating to Paleozoic plants has been derived, often contain mycelium in and about the fragments of roots and other plant organs which they enclose. Records of hyphae in the tissues of plant fossils are also common, and the mycelium occasionally shows characteristic morphological features, e.g. the presence of spherical or pear-shaped swellings or vesicles. Mycelium identical in appearance is associated also with plant fragments in peat deposits of various ages and belongs almost undoubtedly to fungi growing saprophytically in a matrix of decomposing plant residues. Some of the fossil records are believed to point also to the occurrence of parasitic mycelium in plant tissues, and in general, the evidence relating to fossil fungi indicates a mode of life similar to that of the forms now in existence.

In the cortical tissues of *Lepidodendron* or *Stigmaria*, Seward has figured mycelium distributed in the middle lamellae of the cell-walls and forming localised swellings or vesicles within the cells. This recalls the habit observed in endotrophic mycorrhizas at the present day but the material is too scanty to constitute a definite claim for the existence of mycorrhiza in members of this group of fossils. There are, however, at least two records pointing to the formation of mycorrhiza by Paleozoic plants.

Weiss (1904) described a root-like structure containing mycelium from the Lower Coal-Measures under the name of *Mycorrhizinium*. The fossil consisted of fragments of slender roots or possibly leafless rhizomes, the cortical tissues of which provided evidence of differentiation in respect to fungal infection, in itself an indication of mycorrhizal structure.

The distribution of mycelium was mainly intracellular with a local development of spherical and pear-shaped swellings or vesicles, and some evidence that intercellular growth of hyphae also occurred. Even more convincing evidence was provided by the structure of the middle cortical region, many of the cells in which contained dark-coloured structureless masses connected with the surrounding cell walls by radiating strands. The preservation of the tissues was fortunately sufficiently good to permit the identification of these strands as hyphae, and the general structure closely resembled that of corresponding cells containing mycelium in the "clumped" stage of digestion in the endophytic mycorrhizas of *Psilotum* and many other plants (Pl. V, Figs. 36, 37).

Another fossil mycorrhiza was described five or six years later by Osborn (1909) in *Amyelon radicans*, the root of *Cordaites*, the well-known genus of Mesozoic Cycads. It had been observed previously that these roots often showed a development of clustered laterals of stunted growth, and Osborn's discovery that young rootlets from the clusters showed profuse infection of the cortical tissues by mycelium,

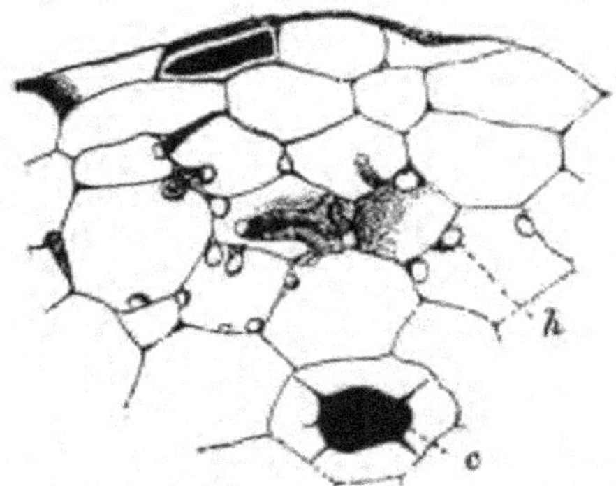

Fig. 48. *Mycorrhiza from the Coal-Measures*. Transverse section showing a portion of the external tissues of a fossil mycorrhiza. A cell of the exo-cortex contains fungal hyphae (*h*), and one of the cells of the medio-cortex a "clump" formation (*c*). (From Weiss, *Ann. of Bot.*, 1904.)

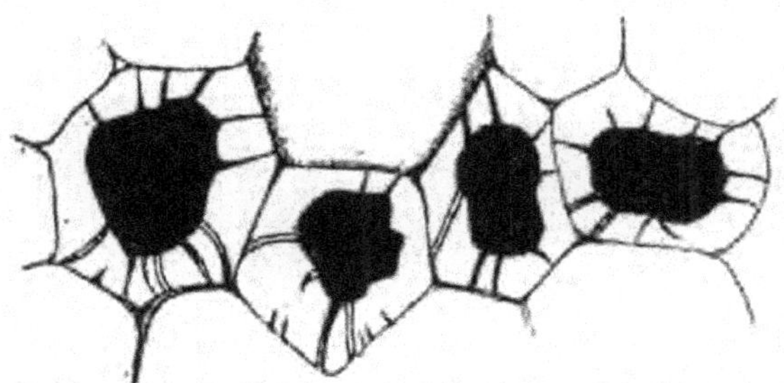

Fig. 49. *Mycorrhiza from the Coal-Measures*. Tangential longitudinal section through the medio-cortex of a fossil mycorrhiza. The cells contain mycelial "clumps" connected by hyphae with the cell-walls. (From Weiss, *Ann. of Bot.*, 1904.)

led him to classify them with the "coralloid" roots so characteristic of living mycorrhiza plants. The localisation of infection to certain tissues and the absence of any apparent injury to the root cells also pointed to their mycorrhizal character. Many cells of the inner cortex were filled with tangled skeins of mycelium, others provided clear evidence of intracellular digestion, while the formation of vesicles was a structural link with the fungi of recent mycorrhizas.

The general relation to the tissues of the hosts shown by the endotrophic mycelium in both these fossils, the evidence for simultaneous existence of intracellular hyphae in an active condition (Pilzwirthzellen) and similar hyphae in the course of digestion by root cells (Verdauungszellen), together with the morphological characters of the endophytes, constitute an irresistible case for regarding them both as endotrophic mycorrhizas of very similar structure to those formed by living plants. If this view is correct, the habit of forming root associations of a harmless or beneficial kind with soil fungi must have been well established in vascular plants as early as the Coal-Measure Period.

Note 1, see p. 119.

Earlier, Busich (1913) had investigated eighteen species belonging to ten genera of Asclepiadaceae of which eleven species showed regular infection, three occasional infection, and four complete freedom from infection whenever examined. Succulent members of the group possessed typical mycorrhiza. In all cases the mycelium was of the usual type with vesicles and arbuscules and suffered extensive intracellular digestion. These observations were believed to support Stahl's view in respect to a direct relation between scanty transpiration and mycorrhiza.

Note 2, see p. 125.

The presence of mycorrhiza in marsh plants has been regarded as exceptional. I am indebted to Miss E. Mason of the Botanical Department, University of Birmingham, for enabling me to confirm her observation that endotrophic mycorrhiza occurs regularly in a number of salt marsh plants, e.g. in *Plantago coronopus* and *P. maritima*, *Aster tripolium*, *Glaux maritima*, and certain grass species. In all cases observed the mycelium was of the familiar "Phycomycete" type subject to intercellular digestion of the usual kind.

CHAPTER VIII

TREE MYCORRHIZA

Ectotrophic mycorrhiza: Müller; Möller; McDougall—Peyronel: the fungi responsible for forming mycorrhiza in trees—Melin: the mycorrhiza of Pine, Spruce, Larch, Aspen and Birch; the ecological and physiological significance of tree mycorrhiza; application of experimental results to field conditions and forestry—Falck's work on the decomposition of humus.

After the publication of Stahl's paper on the subject in 1900, interest in mycorrhiza was transferred temporarily to plants showing the endotrophic type of infection. The first fifteen years of the new century were comparatively rich in researches dealing with the cytology of intracellular infection and saw also the application of new and fruitful methods of experiment to the study of such specialised groups as the Orchids and Heaths. The possible importance of the ectotrophic type of mycorrhiza, characteristic of so many forest trees, as a factor affecting growth continued, however, to attract attention. A classical contribution to the subject was made by Müller (1903 *a*) who published an account of his observations on Spruce and Mountain Pine in Jutland.

Spruce is an important forest tree in the heath region of Jutland, and widespread failure in its culture had attracted the attention of foresters. After a satisfactory start, the trees suffered a severe check, as evidenced by diminished annual growth, decreased size of the needles and dying back of the tops. Whole plantations suffered in this way, either dying off completely or making a slow recovery—in either case causing considerable loss of revenue. Clearing the land of heather before planting and better cultivation caused no improvement, but good results were obtained by planting the Mountain Pine, *Pinus montana*, as a nurse tree in young plantations of Spruce. Müller sought an explanation of this fact and, after examining the factors likely to be operative, fixed upon the mycorrhiza of the Pine as the chief agent responsible for the improvement. He had previously described and figured two forms of mycorrhiza in the Mountain Pine (1902): (*a*) a racemose form with ectotrophic mycorrhiza, and (*b*) a dichotomous form with endotrophic mycorrhiza: in Spruce, on the other hand, he observed only a form with racemose branching and ectotrophic structure. Müller argued that just as Clover and other leguminous species indirectly provide nitrogenous food for other plants by means of their root tubercles, so the Mountain Pine with its endotrophic mycorrhiza, furnished to the Spruce nutrient material inaccessible to the latter when growing alone. Noting that heather and other heath plants also possessed endotrophic mycorrhiza, he inferred that heath conditions were better suited to the Mountain Pine than to Spruce.

If proved, this interesting hypothesis would have an important bearing upon practical forestry. Unfortunately, Müller failed to obtain the requisite experimental proof, while Möller (1906), to whom he supplied seeds and seedlings of *P. montana*, reported that he could find no evidence for fixation of atmospheric nitrogen by the mycorrhizal fungus.

The last-named worker had published two papers (1902, 1903) recording his observations on the growth of Pine (*P. sylvestris*) and Oak, in poor sandy soils and in those containing abundant humus respectively. His attention was attracted by the coralloid mycorrhiza on the roots of Pine (*P. sylvestris*), and he made various attempts to determine its relation to the nutrition of the tree. Frank (1892 *a*, 1894) had previously expressed the opinion that this conifer did not come to maturity on good Pine soils if the formation of mycorrhiza was hindered through absence of the appropriate fungus, whereas it grew vigorously if the roots were infected normally. Möller found experimentally that a direct relation existed between luxuriant development of the trees and abundance of mycorrhiza, but his experiments provided no evidence whatever that the root fungi of either Pine or Oak were able to utilise atmospheric nitrogen. Nor were his efforts to establish the identity of the mycorrhizal fungi more successful, the technique adopted rendering it little probable that the species of *Mucor* isolated from sterilised fragments of root were actually the forms concerned in the formation of mycorrhiza, as indeed was clearly recognised by the investigator himself. He quoted Ramann's experiments (Ramann 1888) which had given similar results, and recorded his own conclusions in the form of an extract from a communication made to him by Sarauw:—"Dass die Pilz unseren Waldbaumwurzeln und den Bäumen Vorteil bringen sollten, ist bisher meines Erachtens weder durch Beobachtungen in der Natur, noch durch Versuche nachgewiesen worden."

In a paper on the mycorrhiza of forest trees, McDougall (1914) described the use of an improved technique for the identification of the fungi present in the mycorrhizas of various trees. It will be recalled that Noack (1889) had put forward a claim for their identity with certain Gasteromycetes and Hymenomycetes commonly found in the neighbourhood of the trees. Noack based his conclusions on the microscopic characters of the mycelia and the frequent association of particular fruit bodies with certain trees. Using the "glass plate" method, McDougall kept individual roots under observation throughout the growing season and, in the paper cited, offered substantial evidence for the identification of the root fungi in six forms of ectotrophic mycorrhiza, viz. *Russula* sp. on *Tilia americana*, *Boletus scaber* var. *fuscus* and *Cortinarius* sp. on a variety of *Betula alba*, and *Scleroderma vulgare* on *Quercus alba*, thus adding four new species to the list of fungi for which claims had been put forward previously. He observed the course of infection and followed the development of the characteristic ectotrophic structure described by Frank, noting the progressive intensity of external infection and its direct correlation with the inhibition of root growth and stimulation of branching which resulted in the development of coral-like clusters of mycorrhiza. McDougall described and figured a number of typical cases and recorded two novel structures, one, a curious type of infected root in Maple with bead-like swellings and a strict localisation of mycelium to the swollen regions, the other a development of what was described as *heterotrophic* mycorrhiza in *Tilia americana*—ectotrophic in general structure, but showing intracellular infection of the cortical cells. He believed that the mycorrhizal relation in trees was a casual one, depending upon the proximity of a suitable fungus or on some chance condition in one or other of the constituents. In respect of the physiology of the relation, he argued that among the endophytic mycorrhizas found in Maples some were associations beneficial to the trees, whereas in others the fungi were simply internal parasites of the roots. The true ectotrophic mycorrhizas of the majority of trees he regarded, without exception, as cases of parasitic attack on the roots, adding in a later paper (1922 *b*):—"It is probable that as a rule no great harm to the higher plant results from this parasitism of its roots by mycorrhizal fungi."

The confusion existing at this time with regard to the identity of the root fungi of trees may be gleaned from the following brief summary: members of each of the great groups, Basidiomycetes, Ascomycetes and Phycomycetes, had been cited by different observers as responsible for mycorrhiza formation. Among Basidiomycetes, Woronin (1885 *a*), Noack (1887) and Frank (1892) had each identified a number of Hymenomycetes as mycorrhizal fungi; Kaufmann (1906), Pennington (1908), Mimura (1915), McDougall (1914, 1922 *b*), Romell (1921), and Peyronel (1920, 1922) added progressively to the list. Boyer (1915–16), tracing the delicate mycelium of species of *Boletus* and *Amanita* through the soil, observed that it could frequently be connected with the mycorrhiza of the tree roots. He believed the same to be true of truffles and other humus-dwelling fungi. Of Gasteromycetes, Noack (1889) had cited *Geaster* sp.; McDougall and Peyronel *Scleroderma vulgare*. Among Ascomycetes the association of truffle fungi with the roots of certain trees had long attracted attention; Mattirolo (1887) and Frank (1892) both believed that species of *Tuber* were mycorrhiza formers; Reess (1880, 1885 *a*) and others had held a similar view for *Elaphomyces granulatus*; Pirotta and Albini (1900) made a new claim for *Terfezia leonis*.

Peklo (1909) contributed an account of his researches on the ectotrophic mycorrhizas of *Carpinus* and *Fagus*, including attempts to extract the root fungi and establish their identity. From roots of both trees he isolated forms of *Penicillium* and *Citromyces* and attempted to prove their identity with the mycorrhizal forms by inoculation of mycorrhiza-free cultures of young Beeches with spore suspensions. The technique adopted by Peklo was somewhat crude, and it is probable that the strains of "Wald-Penicillien" isolated by him were merely members of the profuse epiphytic fungus flora upon the roots used for isolation experiments.

As representatives of the Phycomycetes, Möller (1903) had added the names of several species of *Mucor* extracted from the roots of Pine (*Pinus sylvestris*). In this case, as in the last mentioned, it is unlikely that the forms isolated were other than casual associates of the superficial microflora of the roots. Fuchs (1911) made similar attempts to establish the identity of the root fungi of certain Conifers by the addition of spores of *Fusarium* sp. and *Verticillium* sp. His efforts to reproduce ectotrophic mycorrhiza by these means were unsuccessful. The immediate cutting off of the infected cells observed was interpreted as evidence of parasitic attack on the part of the fungi concerned.

For the inclusion of a majority of the fungi in this rather heterogeneous group, the main evidence consisted of observations on a regular association of particular fruit bodies with the trees concerned; on the continuity of mycelium from the latter with the hyphae of the fungus mantle on the roots; and on the morphological identity of the hyphae attached to sporophores and roots respectively. Although inconclusive, evidence of this nature was cumulative and suggestive in respect to certain Hymenomycetes, more especially in regard to certain species of *Boletus*; the constancy of the records for an association of members of the Tuberaceae with various Cupuliferous trees may also be noted. Certain observers, e.g. Peklo and Möller, had undertaken investigations by means of pure culture methods. In no single case, it may be safely said, had experimental evidence of a satisfactory kind been obtained that the fungus named was actually concerned in the formation of ectotrophic mycorrhiza.

The old problem of the relation of Truffles to forest trees is recalled by certain recent researches on the behaviour of certain species of *Tuber* in pure culture. Matruchot had grown *Tuber melanosporum* and *T. uncinatum* on artificial media and reached the conclusion that these species depended upon an association with Oak for the production of fertile ascocarps. Constantin (1924*a*), on the other hand, has since been successful in obtaining cultures of *T. brumale* which produced both conidia and ascospores on artificial media quite independently of any relation with roots of Oak. The association of fungi of the Truffle series with the mycorrhiza of other trees remains an open question; synthetic cultures recently carried out by Melin with *Elaphomyces* sp. have yielded uncertain results.

The observations of the Italian mycologist Peyronel (1921, '22*b*) on the association of various Basidiomycetes with certain trees have led him to publish the following list as representing almost certain cases of mycorrhizal association.

Fagus sylvatica	*Cortinarius proteus*
	C. bivelus
	Boletus cyanescens
	B. chrysenteron
	Hypochnus cyanescens (nov. sp.)
	Scleroderma vulgare
	Amanita rubens
	Lactarius subdulcis
	L. blennius
	Russula emetica
	R. nigricans
	Cantharellus infundibuliformis
	Hydnum repandum
Corylus avellana	*Lactarius coryli* (nov. sp.)
	L. subdulcis

Corylus avellana	*Boletus chrysenteron*
	Strobilomyces strobilaceus
	Hypochnus cyanescens (nov. sp.)
	Amanita rubens
	Rhodopaxillus nudus
	Cortinarius proteus
	C. multiformis
	C. violaceus
	Hydnum repandum
Betula alba	*Amanita muscaria*
	Amanitopsis vaginata
	Lactarius necator
	L. torminosus
	Boletus scaber
	Scleroderma vulgare
	Russula rhodoxantha (nov. sp.)
	Tricholoma flavobrunneum
Larix decidua	*Amanita muscaria*
	Russula laricina (nov. sp.)
	Hygrophorus bresadolae
	H. lucorum
	Scleroderma vulgare
	Lactarius rufus
	Gomphidius gracilis
	Paxillus lateralis
Populus tremula	*Cortinarius collinitus*
Quercus robur	*Amanita citrina*
	Lactarius subdulcis
	Russula cyanoxantha
Castanea vesca	*Amanita rubens*
	Russula lepida
	R. rubra
	Scleroderma vulgare

Peyronel has pointed out that the general form and dimensions of the various mycorrhizas are determined by the host tree, while the structure, thickness, colour, etc., of the mycelial mantles depend upon the individual fungi present. It is stated, for example, that Larch mycorrhiza formed by *Scleroderma vulgare* is macroscopically indistinguishable from that formed by *Boletus elegans* in the same tree. In certain cases macroscopic characters are said to afford a ready means of identification; thus, the mycorrhiza of beech formed by *Hypochnus cyanescens* is immediately recognisable owing to its fine blue colour. Peyronel has bestowed the new name "mycoclena" upon the fungus mantle, and predicts that it will be possible eventually to make an analytical key which will permit the immediate identification of the fungi present in any mycorrhiza by observing the characters of the mycoclena.

Peyronel has been led to make the new and somewhat remarkable claim that certain Hymenomycetes show constant morphological differences when associated with different trees. For example, the sporophyte of *Boletus scaber* associated with Birch is reported as different from that of the same fungus growing with Hazel, Oak, and Chestnut. Some of the opinions expressed by this author will be received with great caution by mycologists, especially by students of the Agaricineae.

From observations on the west coast of Sweden, Romell (1921) was struck by the constant presence of "fairy rings" of the "Butter Fungus," *Boletus luteus* around *Pinus montana*, independently of soil differences. A similar association had been observed elsewhere, not only with the Mountain Pine, but also with *P. austriaca* and *P. sylvestris*. Earlier, Quélet had reported *Boletus boudieri* in association with *Pinus halepensis* and *Pinus pinaster*, and *B. elegans* with *Larix*. Attempts to trace a direct connection of the mycelium of *Boletus luteus* with the roots of *P. montana* did not yield conclusive evidence, but Romell's conviction that this existed, as also his view in respect to the significance of the association between *Boletus elegans* and Larch, have been fully justified by subsequent researches.

The close association of a "fairy ring" of Hymenomycete sporophores with a given species of tree occasionally observed in nature is illustrated in the photograph reproduced in Fig. 52, although in this case the identity of the species of *Clitocybe* to which the fructification belongs with any of the fungi present in the mycorrhiza of the tree was not investigated. (Pl. VI, Fig. 52.)

In a brief account of mycorrhiza in Sitka Spruce (*Picea sitchensis*) and other trees, Laing (1923) has supplied an analytical table of the various types of mycorrhiza observed by him in conifers and noted that both the ectotrophic and endotrophic types of structure are represented as well as an intermediate or "semi-ectotrophic" type. *Thuja gigantea*, *Sciadopitys verticillata*, and *Pinus sylvestris* are cited as forms showing intracellular infection. In discussing the unsatisfactory character of the evidence relating the mycorrhiza of trees to the intake of food material, Laing describes certain observations of his own indicating the presence of relatively large amounts of oxidising enzymes—oxidases and peroxidases—in the mycorrhiza of trees, and discusses the possible advantage of this increased oxidising mechanism to the trees in habitats of deficient aeration and available plant food. In an account of field observations in south-east England, Paulson (1924) has recently drawn attention to the profuse development of tree mycorrhizas in the superficial rooting system that does not penetrate the soil but ramifies throughout the accumulations of fallen leaves near the trees. This distribution, if general, may possess some significance in relation to Laing's observations on the presence of oxidases. (Pl. VI, Fig. 51.)

Of other coniferous mycorrhizas, that of *Abies firma* was described by Tubeuf (1896), Noelle (1910), and Mimura (1917), none of whom put forward any views as to the identity of the fungus or fungi responsible. More recently, Koki Masui (1926) has observed sporophores of *Cantharellus floccosus* attached to the mycorrhizas of this conifer, and so added another species to the growing list of Basidiomycetes believed to be responsible for mycorrhiza formation. The Japanese observer distinguishes four distinct types of mycorrhiza in this Fir, one being an ectotrophic form with curious basidia-like projections from the mantle, an intercellular "Hartig net" of the usual kind, and more or less profuse intracellular infection. The mycorrhizas associated with *Cantharellus* were found only in the superficial layer of soil, and the fungus was observed to occur only in dry situations. Infection reduced the rate of growth of the roots, and was sometimes fatal. The author agrees with McDougall in regarding the relation as essentially one of parasitism on the part of the root fungus. In a later paper (Masui 1926 *b*), he has described and figured the renewed growth observed in these and similar mycorrhizas after the formation of the mantle. The mycelial sheath splits in several directions and a new fungal investment develops from the margins of the fissures.

Of recent attempts to identify the mycorrhizal fungi of trees by isolation methods may be mentioned that of Chan (1923), who isolated a non-sporing mycelium from roots of Beech growing in Munich and in Tyrol and named it *Mycelium radicis fagi*. It is reported that under certain conditions of nutrition—e.g. the presence of one per cent. peptone in the culture medium—the hyphae of this form become shortly jointed like those of the Orchid fungi. On various grounds, the mycelium was believed to be that of the mycorrhizal fungus of Beech, but no experimental proof of its identity has yet been obtained.

In respect to the significance of mycorrhiza in trees, it may be recalled that every imaginable view has found supporters.

Gibelli (1883), Sarauw (1893), Möller (1903), Pastana (1907), Ducomet (1909), Fuchs (1911), McDougall (1922) and Koki Masui (1926) have regarded all or certain of the root fungi as more or less mischievous parasites. Henschel (1887) reached the conclusion that all Spruces developing unhealthy symptoms possessed mycorrhiza, that the degree of damage to the tree was proportional to the amount of infection, and that the strongest saplings were fungus-free. He concluded that the effect was without exception damaging to the young plants:—"Dass die Einfluss dieses Symbioten auf der Entwicklung der jungen Fichtenpflanze als ein absolut schädlicher angesehen werden müsste." On the other hand, Frank (1892), Stahl (1900), Müller (1903), Tubeuf (1903), Elenkin (1907) all believed the association to be one of beneficial symbiosis.

Efforts have been made to study the physiological relations by chemical and microchemical methods. Weyland (1912) made a serious attempt to review the evidence bearing on the nutrition of mycotrophic plants in general, and was the first observer to use microchemical methods for studying the distribution of inorganic nutrients—phosphorus, potassium and calcium—in the root cells. From his own observations on the roots of autotrophic and mycotrophic plants he believed that these methods offered the most promising line of research for gaining new information as to the nutritive relations of members of the latter group with their fungal associates. His conclusions with respect to ectotrophic mycorrhiza agreed with those expressed by Fuchs (1911), a contemporary observer, viz. that the root fungi of trees were parasitic upon their hosts and that no symbiosis of a kind beneficial to the latter existed.

Weevers (1916), examining the distribution of ammonium salts, discovered that these were present in very small amounts or were entirely lacking in mycorrhiza plants. Hence, he argued, assimilation of nitrogen, if facilitated by the root fungi, must be brought about in a different manner from that in leguminous tubercles, in which these salts are present in quantity. More recently, Rexhausen (1920) carried out observations on a number of ectotrophic mycorrhizas by similar methods. Using as material species of *Pinus*, *Quercus sessiflora* and *Monotropa hypopitys*, growing in the open, he reached the conclusion that the root fungi facilitated the absorption of mineral salts, i.e. he found evidence in support of the Stahl theory of nutrition. Rexhausen also expressed the opinion that the relation between fungus and root in ectotrophic mycorrhiza was not a fixed one, but fluctuated with the soil conditions. If these were unfavourable to the fungus, it might parasitise and greatly injure the tree, if they were too favourable, infection was feeble and mycorrhiza badly developed. On the whole, in the coniferous trees studied, he found evidence that a symbiosis of a beneficial kind existed, especially under soil conditions such that a supply of carbohydrates outside the roots was associated with a deficiency of mineral salts. His observations upon *Monotropa* did not throw any fresh light upon the problem of nutrition in this species.

The recent researches of Elias Melin (1917–1925) mark a notable advance in the knowledge of ectotrophic mycorrhiza. By the publication of this series of papers the author has provided, for the first time, reliable and conclusive evidence of the nature and constitution of the mycorrhizas of certain trees and the identity of the fungi concerned. The work is of such interest and importance that its consideration at some length calls for no apology.

In a brief preliminary note on the mycorrhiza of Scotch Fir and Spruce, Melin offered the following new facts for consideration. (*a*) The fungi present in ectotrophic mycorrhiza are not strictly intercellular in distribution as was claimed by Frank; the mycelium invades many of the cortical cells and can be put in evidence by suitable staining methods. (*b*) In addition to the normal ectotrophic type of mycorrhiza in Pine and Spruce, there appears on the roots of individual trees growing in wet bogs and occasionally also in raw humus elsewhere, a fungus association of a different character in which the "Hartig net" and mycelium mantle are lacking and the mycelium is exclusively intracellular. (*c*) Two fungi concerned in the formation of the mycorrhiza of Pine and Spruce have been isolated and their identity established by means of inoculation cultures; provisionally they were named *Mycelium radicis silvestris* and *Mycelium radicis abietis* (Melin 1922*a*). A full account of these important researches was published subsequently in the form of a monograph and will now be considered in greater detail. (Melin, 1923 *a*.)

The more significant results fall naturally under three headings:—

(1) An account of mycorrhiza in Pine and Spruce as it occurs in nature.

(2) The isolation, cultivation and description of the root-fungi; and the reconstitution of the mycorrhizas under pure culture conditions.

(3) The experimental identification of the mycelium of certain root-fungi with Hymenomycete fructifications associated regularly with the trees in nature.

A. The Mycorrhiza of Pine (*Pinus sylvestris*) and Spruce (*Picea abies*).

In conifers, as in most deciduous trees, it is the actively absorbing roots which become mycorrhizas. The differentiation of the root system into "long" and "short" roots is thereby greatly accentuated, the distinction between the two types being much less when they are protected from fungal infection. In Sweden, mycorrhiza is formed in spring and autumn, the periodicity corresponding with that of root growth showing two maxima, in spring and autumn respectively. Several types have been observed in Pine and Spruce, differing in mode of branching, colour and internal structure.

For the former tree, Melin has described three forms:—"Gabelmykorrhiza," "Knollenmykorrhiza" and simple mycorrhiza ("einfach Mykorrhiza"). The first type, noted and recorded by many earlier workers, is the commonest; it is best developed in woodland soils with an abundant layer of raw humus, but is found also on moorland soils, especially after drainage. The colour varies from golden brown to black while other tints have been observed, the colour differences depending upon the fungi present. The habit of this type shows well the arrest of growth and crowded dichotomous branching so often described, leading to the production of a dense tuft or "witches broom" of characteristic appearance. According to Melin, a majority of root-fungi form this type of mycorrhiza in nature (Pl. VI, Fig. 50).

"Knollenmykorrhiza" of Pine is also abundant in Swedish woods on similar types of woodland soil. It presents the appearance of small tubers, variable in size and often so crowded that they have grown together. The tuberous habit is due to the merging of a cluster of dichotomously forked roots into one structure by fusion of the individual fungus mantles. The colour, at first pale, becomes grey to brownish grey with age. The surface of the tubers is rough, owing to the passage outwards from the mantle of strands of mycelium that can be traced into the humus, where they break up into separate hyphae (Fig. 50). Not the least interesting aspect of Melin's work has been the definite association of this type of mycorrhiza with the species of *Boletus* responsible for its formation, viz., *B. luteus*, *B. granulatus*, *B. variegatus* and *B. badius*. He concluded it to be identical with the forms described by Müller (1903) for *Pinus montana* in Jutland and by McDougall (1922) for *P. strobus* in the United States. Both this and the preceding type may occur in close proximity on the same main root (Pl. VI, Fig. 50).

The simple mycorrhizas of the Pine are frequently young stages of one or other of the types just described. Sometimes, however, they are believed to represent a special type, in which case they are longer (10 mm. as compared with 4 mm.) and thinner (0·2 mm. as compared with 0·4 mm.). These unbranched mycorrhizas are characteristic of Pine heaths and their formation is believed to depend upon a decreased "virulence" of the root fungi under the conditions existing in heath soils.

Two types of mycorrhiza in Spruce (*Picea abies* L.) have been described; the racemose (razemose) and the simple (einfach) (cf. Müller, p. 339). The former is the characteristic form under favourable soil conditions. The branching is of monopodial type with lateral roots in two rows upon a main axis, fungus infection being similar in the roots of both orders; free branching may produce a cluster of short roots very different in appearance from the typical "Gabelgebüsche" of Pine. As in the latter, the colour is variable, with often a close association of different colours on the same root. The simple type of mycorrhiza resembles that of Pine.

The development and structure of the fungus mantle in these mycorrhizas of Pine and Spruce were found to be very variable, a fact afterwards correlated with soil conditions and the presence of different fungus constituents. For example, in the *Boletus* mycorrhizas, a thicker mantle was formed than with other root fungi, and, in general, the mantle is better developed in the raw humus of Pine woods than on heath soils. The significance of these observations will be discussed hereafter.

The histological characters of the ectotrophic mycorrhizas of conifers have been variously recorded as agreeing with those originally described by Frank (1885) with fungus mantle and intercellular "Hartig net," or showing a truly endotrophic structure as described by Sarauw (1893), or of ectotrophic type with a profuse development of intracellular mycelium as recorded by Peklo (1913) for both Pine and Spruce. These inconsistencies are ascribed by Melin to two causes,—faulty technique on the one hand, and the existence of actual differences correlated with the presence of different root fungi and (or) variation in soil conditions, on the other hand. According to the last named author's observations two kinds of "Gabelmykorrhiza" occur in Pine, one of true ectotrophic type, the other with heavy endotrophic infection of the cortical cells. Noteworthy features in the former are the development of a conspicuous "tannin layer" from the epidermal and sub-epidermal layers of the root, the cells of which are constantly penetrated by fine hyphae, as well as the presence of a granular layer (Körnerschichte), comprising two or three layers of cortical cells filled with granules of varying size and surrounded by the pseudotissue of the "Hartig net" but otherwise free from fungal infection. In the other form of this mycorrhiza, the "tannin layer" is less conspicuous, and the granular layers are replaced by a region of profuse intracellular infection in which the mycelial contents undergo rapid and complete digestion, the cytological features of the digesting cells resembling those found in Orchids in respect to increased size of the nuclei and the ultimate removal of the stainable products of digestion.

Both types of mycorrhiza, especially when old, are subject to structural modification owing to the attack of a pathogenic fungus, subsequently isolated and named *Mycelium radicis atrovirens.* It is suggested by Melin that the parasitic hyphae of this pathogen were mistaken by Peklo (1913) for those of the true symbiont, and also that the same mycelium was erroneously described as belonging to species of *Cladosporium* by this author.

The "Knollenmykorrhiza" of Pine

The individual roots of the tuberous complexes formed in this type of mycorrhiza are distinguished structurally by the large size of the "cells" of the pseudoparenchyma of the mantle. They are also remarkable for the profuse and typical endotrophic infection of a five- or six-layered cortical region, each cell of which shows active digestion of the intracellular mycelial complex with subsequent disappearance of the stainable products. Without discussing the anatomical evidence in greater detail, it may be stated that Melin regards the structure of all three types of mycorrhiza in Pine as indicating an exchange of food material and hence, pointing to the conclusion that they constitute a working case of mutual symbiosis—"Wurzel und Pilz leben in mutualistischer Symbiose." The microchemical evidence obtained by Rexhausen (1920) is cited in support of this view. A critical consideration of its correctness or otherwise may be postponed to a later chapter in view of the recent publication of the results of an experimental enquiry into the same matter. (Melin 1925 *c*.)

The Mycorrhiza of Spruce (Picea abies *L.*).

Both the structural types of "Gabelmykorrhiza" found in Pine are represented in Spruce, but there is no counterpart in the latter of the "Knollenmykorrhiza" just described. The mycelium of the true symbionts occurs both in the long and short roots of Spruce but was not found in the older main roots. As in Pine, the mycorrhiza of Spruce is subject to the attack of a parasitic fungus which was identified later with that named *Mycelium radicis atrovirens* by the author.

From the foregoing brief account it is apparent that two distinct anatomical types of mycorrhiza are formed in conifers. Melin questioned the propriety of including those showing both intra- and inter-cellular infection with the true ectotrophic forms, and in view of the fact that the former combine the characters of ectotrophic and endotrophic types, he named them *ectendotrophic* (*ehtendotrophic*). Mycorrhiza of this composite type was known to occur also in *Larix* (Melin, 1922 *b*), and *Betula* (Melin, 1923 *b*), and from the literature may be assumed likewise in *Pinus cembra* (Tubeuf, 1903), and *Tilia americana* (McDougall, 1914).

Comparing it with that described by McDougall in *Tilia*, the designation *heterotrophic* used by the latter author was rejected by Melin on the grounds that the word has a very different significance in other connections. It was probably mycorrhiza of this type that Peklo (1913) described for Pine and Spruce in Bohemia, but Melin could find no evidence that a species of *Penicillium* is concerned in its formation, as believed by the former observer.

Mycelium was not observed in the aerial parts of either Pine or Spruce, and attempted isolations from the shoots gave negative results.

Pseudomycorrhiza.

The existence of a "false" mycorrhiza in both Pine and Spruce had been previously recorded (Melin, 1917), the fungus constituent being regarded in general as a one-sided parasite, although the relation fluctuated with soil conditions and might result either in defeat of the fungus by unusually active digestion or in death of the invaded cells. These pseudomycorrhizas are usually simple and unbranched and in Sweden are specially characteristic of moorland soils afforested after draining. Their recognition offers a reasonable explanation for inconsistencies in the records of other observers, e.g. those of Möller in Brandenburg. It is not improbable that the types of root-infection recorded by Möller from this district were pseudomycorrhizas of similar character.

The Mycorrhizal Fungi of Pine and Spruce and the Synthesis of Mycorrhiza.

Isolation of the fungi responsible for forming the types of mycorrhiza just described, with proof of their identity by the synthesis of mycorrhiza under "pure culture" conditions, constitute an essential part of Melin's researches.

A constant association of the mycorrhizas of Conifers with the fruit bodies of certain Basidiomycetes had been reported by numerous observers, although experimental proof of the continuity of the two systems of mycelia was lacking in every case. The long and heterogeneous list of fungi named in this connection has already been noted and the pressing need, as Melin quickly recognised, was for synthetic cultures on similar lines to those successfully used in the experimental work on *Ericaceae* and the Orchids. Two variations of such methods are available: one, the isolation of the appropriate fungi from roots and the re-synthesis of mycorrhiza under "pure culture" conditions; the other, the formation of mycorrhiza under such conditions by inoculation from pure cultures of known fungi suspected of symbiotic relations. Both methods have been successfully used by Melin who thus obtained corroborative proof for the association of certain well-known Basidiomycetes with the roots of coniferous trees.

With regard to the technique adopted, it may be noted that satisfactory methods were devised for the external sterilisation of roots, and for the exclusion of fast-growing forms, e.g. species of *Penicillium, Mucor, Fusarium*, etc., likely to outgrow the true endophytes. For the detailed characters of the latter and their behaviour in pure cultures, reference must be made to the original records. Their essential features are indicated in the following brief summary.

(*a*) *The mycorrhizal fungi of the* "Knollenmykorrhiza" *of Pine.* This type of mycorrhiza from many different stations yielded a number of fungus forms of similar type. Owing to the comparative ease of external sterilisation of the tubers and the relatively vigorous growth of the mycelia, their isolation was a comparatively simple matter. The characters and behaviour of all the strains in pure culture pointed to their inclusion in a single genus. Two morphological peculiarities may be noted: a characteristic paired branching of the mycelium in young cultures, and the invariable presence of "clamp connections" (Schnallen) joining contiguous "cells." The first-named character appeared also in mycelium associated with the outside of the tubers and was later found to be characteristic of that of certain species of *Boletus* frequent in Pine woods; the last-named is characteristic of many, but not of all Hymenomycetes (Kniep, 1915); it served in this case for the classification of the endophytes into two groups characterised respectively by few or numerous 'clamp connections.' To the whole series of forms was given the name *Mycelium radicis silvestris α.* Later work supplied the necessary proof that the mycelium known by this name belonged to various species of *Boletus* abundant in Pine woods. Of these, *B. granulatus*, *B. variegatus*, *B. luteus* and *B. badius* have been identified with certainty as constituents of the tuberous type of Pine

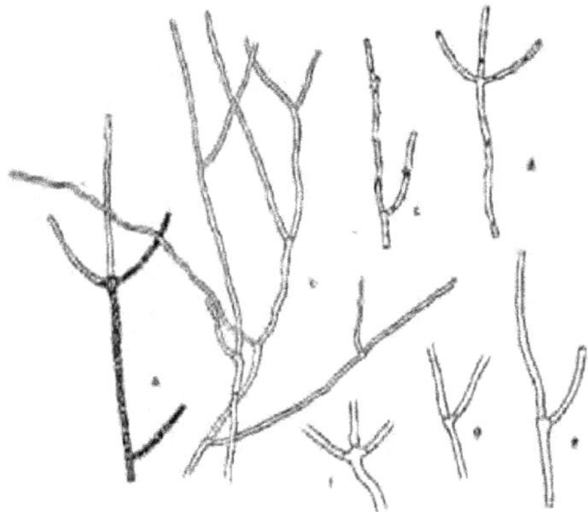

Fig. 53. Aerial hyphae of *Mycelium radicis silvestris α* in pure culture. Strains isolated from trees in different parts of Sweden. (From Melin, 1925 *a*.)

mycorrhiza. *B. bovinus*, a common constituent of the fungus flora of Pine woods, may probably also be included.

The characteristic mode of branching exhibited by the mycelium of *M. r. silvestris α* in common with members of the genus *Boletus* is shown in Fig. 53. It is of interest to compare with this the mycelium figured by Rylands in 1844 as a constituent of the mycorrhiza of *Monotropa hypopitys* (see Fig. 2). The association of these forked hyphae with the roots of *Monotropa* may have been quite fortuitous; it may, on the other hand, be regarded as a possible clue to the identity of the root fungi of this species, in respect to which nothing is yet known.

(*b*) *The mycorrhizal fungi of the* "Gabelmykorrhiza" *of Pine.* Much greater difficulty was experienced in isolating the fungi responsible for this type of mycorrhiza, owing to their poor growth in artificial media. Three forms, differing in the morphological characters of their mycelia and believed to be representatives of a much larger group, were isolated and named respectively *Mycelium radicis silvestris* β, *Mycelium radicis silvestris* γ, and *Mycelium radicis silvestris* δ. The last-named was characterised by particularly feeble growth in artificial cultures. The mycelium of these strains showed Hymenomycete characters. Important differences in their structure and behaviour in pure culture rendered it unlikely that they belonged to a single genus or were nearly related as was the case with the forms included under *Mycelium radicis silvestris* α. Examination of the mycelium of various Hymenomycetes common in Pine woods provided no positive evidence of identity, although that of a number of species of *Tricholoma* and *Cortinarius* showed features resembling those in *M. r. s.* β, and *M. r. s.* γ respectively. No evidence was forthcoming for the identity of *M. r. s.* δ. Further researches revealed the fact that several of the forms isolated could form mycorrhiza on the same tree; whether they might also be present in the same mycorrhiza was not ascertained.

In addition to the true root fungi, a form referred provisionally to the genus *Rhizoctonia* under the title *R. silvestris* was isolated from roots of young Pines which showed unsatisfactory growth. This fungus did not form typical mycorrhiza and was regarded by Melin as a parasite, sharply distinguished from the specific mycorrhiza formers.

(*c*) *The mycorrhizal fungi of Spruce.* Owing to their extremely feeble growth in pure culture, the mycorrhizal fungi of Spruce were likewise difficult to isolate, and only one form, *Mycelium radicis abietis*, was extracted and cultured. The mycelium showed Hymenomycete characters recalling those of *M. r. silvestris* γ from Pine, and it was believed to belong to a member of the same unidentified genus.

Final proof of the identity of the fungal forms just described with those responsible for forming mycorrhiza depended upon synthesis of the mycorrhizas under pure culture conditions. To this end, seedlings of Pine and Spruce were raised from sterilised seed and grown in small flasks in sand or humus cultures. The experimental seedlings developed normally without infection and showed certain structural features of interest in addition to the absence of mycelium, e.g. the absence of any marked differentiation to "long" and "short" roots, as well as a profuse development of root hairs from the main roots and laterals. Inoculation experiments proved that each of the forms included under the names *Mycelium radicis silvestris* α, β, γ and δ from Pine, and *M. r. abietis* from Spruce, formed mycorrhiza. Owing to the fact that, even after leaching, sterilised humus is extremely toxic to these root fungi, sand, watered with a suitable nutrient solution, was found to be more favourable than humus for pure culture experiments, and it was only in such sand cultures that synthetic mycorrhizas were formed.

In general, the root fungi showed a marked increase in vigour when brought from pure culture into contact with the roots of ex-

Fig. 54. Synthetic mycorrhizas of *Pinus montana* produced in pure culture with *Mycelium radicis silvestris* α. k. Knollenmycorrhiza. × about 2½ times. (From Melin, 1925 *a*.)

perimental seedlings. Although more vigorous than when growing alone, they were held to be less so than in nature, as evidenced by e.g. the absence of sporophores. Suggested causes for such decreased vigour were the absence of specific food substances normally present in natural humus, and the accumulation of the toxic by-products of growth in the culture flasks. The latter condition was known to have a depressing effect upon growth in pure cultures.

Synthetic mycorrhizas of two kinds were observed in these and similar cultures, some of the type described as pseudomycorrhiza with intracellular infection only, others of the *ectendotrophic* type described above. Some of the latter exactly resembled those found in nature, others lacked the mycelial investment or mantle; in general, they were unbranched, but both the "Gabel"- and "Knollen"-mycorrhizal types were observed in artificial cultures. The details of structure, e.g. development of an intercellular "Hartig net" and the occurrence of intracellular infection with subsequent digestion, were similar to those in natural mycorrhizas (Fig. 54). It is noteworthy that artificial mycorrhizas of the pure ectotrophic type were not observed.

From comparative observations on Pine, Spruce and Larch, Melin subsequently reached the conclusion that the different types of mycorrhiza observed in nature represent phases in the development of infection, the typical ectotrophic condition being a final stage in a gradual "squeezing out" process due to enzyme activity in the root cells. On this view the endotrophic structure sometimes observed is transitional to the ectendotrophic condition, which in turn may give rise to true ectotrophic mycorrhiza with a thick mantle, well-developed intercellular net, and rare and intermittent intracellular infection. Fungal strains weakened by long-continued growth on artificial media or grown under unfavourable conditions, e.g. in sterilized humus or on a substratum of unsuitable H-ion concentration, can cause only endophytic infection, the exact nature of which will vary with the vigour or "virulence" of the strain used; the more vigorous strains can produce an intercellular net together with a well-developed external mycelial sheath or mantle. It was admitted that the absence of the ectotrophic type of structure from artificial cultures might be explained otherwise, for example, by assuming that the fungi responsible were more difficult to isolate, and that those actually extracted were the forms concerned in forming the ectendotrophic type, but it is clear that Melin himself inclines to the former view. The vigorous branching, so characteristic of natural mycorrhizas, was believed to be a direct consequence of the ectotrophic habit, and therefore, accepting the view just stated, to depend upon the presence of a fungus strain of relatively high "virulence." The higher the "virulence" the more marked the effect upon branching. Hence, under artificial conditions, one and the same fungus might give rise to *simple* mycorrhiza, "Gabelmykorrhiza," or "Knollenmykorrhiza" according to the degree of "virulence" induced by the conditions in respective cultures. The vigour or "virulence" of the individual strains was believed to be bound up with soil conditions nutritive and otherwise. This aspect of infection has since been more fully investigated and dealt with by the author in a separate communication (Melin, 1925 *c*). (See p. 159.)

It has been shown by experiment that certain of the mycorrhiza-forming fungi are more specialised than others: some can form mycorrhiza in members of several genera, others only in those belonging to a single genus, or, in extreme cases, in only a few species of one genus. To the more specialised group belongs *Boletus elegans*, a mycorrhizal fungus of Larch (Melin, 1922 *b*), to the less specialised, other species of *Boletus*, e.g. *B. scaber* and *B. rufus*, mycorrhiza formers in Birch and Aspen. It is still unknown whether the forms associated with Pine and Spruce belong to the more specialised types. It was, however, observed in synthetic cultures, that not all those isolated could form mycorrhiza in a particular tree with the same ease, for example, in Spruce, *M. r. abietis*, *M. r. silvestris* β and *M. r. s.* γ are extremely active, while *M. r. s.* α (*Boletus* spp.) have a relatively feeble capacity to form mycorrhiza, i.e. they are mycorrhiza formers of the second order. (See p. 158.)

In addition to the true mycorrhizal fungi, it has been ascertained by Melin that a large number of casual soil species can cause root infection of the pseudomycorrhizal type. Certain of these casual associates, e.g. *Penicillium* sp. (see Peklo, 1909), and *Acrostalagmus* sp. are harmless to the trees; others, e.g. *Mucor ramannianus* (Möller, 1903) and *Verticillium* sp. (Fuchs, 1911 *a*), behave as parasites. The mycelium of the fungus named *Mycelium radicis atrovirens* was frequently found to be present in old and young mycorrhizas, especially in the former, and proved comparatively easy to isolate. This mycelium formed sclerotia in culture and within the root cells, and it is believed that the species has affinities with the genus *Rhizoctonia*. Under pure culture conditions the fungus was purely parasitic and behaved as a pathogen; as observed by Melin, its presence in natural mycorrhizas is probably deleterious to growth and hinders the action of the true root fungi. It is possibly identical with the form in the mycorrhiza of Beech described by Peklo (1913) as:—"Kein Mykorrhizenbildner, sondern Parasit sei."

The Experimental Identification of Certain Hymenomycetes with the Mycorrhizal Fungi of Pine.

In an earlier paper Melin (1922 *c*) had stated his conviction that the mycelium of the root fungi of Pine included under the name *Mycelium radicis silvestris* α belonged to the genus *Boletus* —"*M. r. silvestris* α gehört zur Gattung *Boletus*, umfasst aber wahrscheinlich mehrere Arten, von denen *B. luteus* eine ist." This belief, based on close resemblances in the mycelial characters and the frequent association of sporophores with the trees, was tested experimentally by the inoculation of pure culture seedlings from cultures isolated from the sporophores of various species of *Boletus* found growing naturally. The forms investigated, *B. luteus*, *B. variegatus*, *B. granulatus* and *B. badius*, are practically confined to coniferous woods and are specially characteristic of pure pine woods or mixed woods of Pine and Spruce. They were easily isolated and grew readily on malt agar and malt gelatine.

Synthetic cultures showed that all four species formed mycorrhizas in Pine. Under culture conditions only simple or dichotomously branched mycorrhizas (Gabelmykorrhiza) were formed, but it was believed that, under the more favourable conditions existing in natural humus, similar root infection led also to the formation of the tuberous type of mycorrhiza (Knollenmykorrhiza). In woods it is usual to find the latter strongly developed in the humus below fruit bodies of *Boletus* sp. Certain characteristic features of natural "Knollenmykorrhiza," e.g. the large-celled character of the mantle, were also noticeable in the synthetic mycorrhizas.

From analogy with Birch (Melin, 1923 *b*) it was regarded as likely that species of *Amanita* and *Tricholoma*, genera well-represented in Pine and Spruce woods, were also concerned in the formation of coniferous mycorrhiza. It may be recalled that the mycelium of the form *M. radicis silvestris* β, closely resembled that of species of *Tricholoma*.

B. The Mycorrhiza of *Pinus montana*.

The conclusions reached by Müller (1903) in his historic investigation on the relation of the Mountain Pine to the growth of Spruce on the Jutland heaths (p. 129) rendered the root biology of the former tree of special interest to botanists and foresters. In order to ascertain whether the same mycorrhizal fungi were present as in *P. sylvestris*, Melin extended his investigation to include a series of pure culture experiments with seedlings of *P. montana*. They proved more amenable to culture under experimental conditions than had those of *P. sylvestris*, and the synthetic mycorrhizas developed in sand cultures more closely resembled those formed in nature. The production of pseudomycorrhizas in certain cases was attributed to the use of fungus cultures weakened by long cultivation, and consequently lacking the "virulence" necessary for the formation of the typical condition (Fig. 54). Using three of the fungus strains isolated from *P. sylvestris*, viz. *M. radicis silvestris* α, β, and γ, it was found that all these forms likewise formed mycorrhiza in *P. montana*. In the *a* group (*Boletus* spp.), it was proved by direct experiment that *B. granulatus*, *B. luteus* and *B. variegatus* also formed mycorrhiza in Mountain Pine; that *B. badius* did not form typical mycorrhiza in the latter in synthetic cultures was attributed to the marked tendency shown by this species to degenerate under "pure culture" conditions.

A number of other Hymenomycetes common in Swedish woods were also tested experimentally, and it was determined that *Russula fragilis*, *Lactarius deliciosus*, *Cortinarius muscosus* and *Tricholoma virgata* all formed synthetic mycorrhizas in *P. montana*. Furthermore, it was regarded as probable that other species of these genera, and also *Amanita rubescens* would give similar results if tested in like manner. In general, it was concluded that the two species of Pine possess the same mycorrhizal fungi (1924 *a*).

C. The Mycorrhiza of Larch.

Following his own observations and experiments, Melin (1922 *b*, 1925 *a*) reached the conclusion that an obligate mycorrhizal relation with the mycelium of *Boletus edulis* existed in Larch. Subsequently, Hammerlund (1923) reported that he was unable to obtain experimental confirmation of this and he rejected Melin's conclusions on the subject. In his most recent contribution to the literature of mycorrhiza, Melin (1925 *c*) has restated his former conclusions, citing *B. elegans* as an example of the most specialised type of mycorrhizal association, in which a given fungus forms mycorrhiza only with members of a single genus of vascular plants:—"Zu den spezialisiertesten gehört *Boletus elegans*, der ganz und gar an die Lärche gebunden zu sein scheint."

The statement that this species of *Boletus* is associated with trees other than Larch in southern Europe (Lange, 1923) is believed to require confirmation.

D. The Mycorrhiza of Birch and Aspen.

In order to acquire a many-sided acquaintance with tree mycorrhiza, Melin made a study of Birch (*Betula pendula* Roth. and *B. alba* Roth.), and Aspen (*Populus tremula* L.) as representatives of deciduous trees (Melin, 1923). As in the case of coniferous mycorrhiza, there exists an extensive and contradictory literature on the subject; certain Hymenomycetes, regularly found in Birch woods, had been repeatedly named as mycorrhiza formers, but experimental evidence of their identity was in every case lacking. Paulson (1923) has published an account of Birch mycorrhiza and described the infection of the styles of the ripe fruit by mycelium of a fungus believed to be *Sporotrichum pulviniforme*. The germinating seedling is exposed to infection by these hyphae, which, it is suggested, may be causally related to the subsequent formation of mycorrhiza. There is at present no experimental support whatever for the correctness of this interesting suggestion.

Melin's observations on Birch and Aspen have established the main facts experimentally, and have shown incidentally that the conflicting observations of earlier workers with regard to structure was due to faulty technique. The mycorrhizas of these trees are simply or monopodially branched, yellow or yellowish brown when young, becoming dark brown to black with age. They are of the ectendotrophic type with profuse intracellular infection. The strictly ectotrophic structure reported by Mangin (1910) and McDougall (1914) was not observed in Sweden, a discrepancy explicable either by non-formation of the latter under certain conditions of soil and climate, or by assuming a defective technique on the part of the earlier observers. A similar type of mycorrhiza is formed by both trees; since it differs in several respects from the coniferous type, that of Birch will be briefly described.

In transverse section the mycorrhiza of Birch shows a mycelial mantle of the usual ectotrophic type, a palisade layer of large, radially elongated cells with rich intracellular infection, and an inner zone several cell layers thick with profuse infection and typical intracellular digestion. The cells of the palisade layer are separated by a small-celled intercellular "net" of pseudoparenchyma, and contain hyphae of two kinds, described respectively as protein hyphae (*Eiweisshyphen*) and haustorial hyphae (*Haustorienhyphen*). The former are of large diameter (10 μ), with abundant protein contents and very large and conspicuous nuclei. They are branches from hyphae in the inner zone of the mantle which, traversing the palisade cells, may branch laterally, and eventually penetrate the cells of the digestive layer. The haustorial hyphae are of small diameter, with scanty contents and very small (*winzig*) nuclei; they may fragment in the palisade cells or penetrate the cells of the innermost zone where they undergo complete disintegration and digestion with subsequent disappearance of the stainable products (Fig. 55). As is usual in mycorrhizas, the endodermis and vascular cylinder is free from infection.

From structure alone, Melin has established a good case for the exchange of food material in these mycorrhizas with resulting benefit to the trees:—"Der anatomische Bau (des Birkes) zeigt dass die höhere Symbiont von den Pilzhyphen keineswegs geschädigt wird." The evidence for this, as for a similar condition in coniferous mycorrhiza, will be considered later. As well as the typical mycorrhizas, pseudomycorrhiza of the kind described for Pine is formed both by Birch and Aspen.

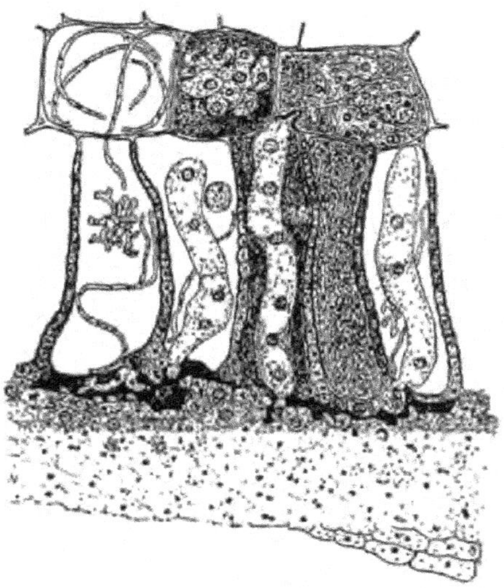

Fig. 55. Longitudinal section through a Birch mycorrhiza. Original drawing × 1000. Reduced about ⅓. (From Melin, 1923 *a*.)

As regards the identity of the root fungi, the commonest Hymenomycetes in northern Birch woods are *Boletus scaber* and *Boletus rufus*. The presence of these species is practically invariable but has been recorded also in other deciduous woods and exceptionally in coniferous woods. Thus Thesleff (1919) observed both species in Larch, Oak, and Alder woods; Peyronel (1917) reported *B. scaber* under *Fagus*, *Corylus*, *Castanea*, and *Sorbus aucuparia*. Moreover, Fries (1874) had noted the association of *B. scaber* with Birch and Köppen (1889) had recorded it from Birch woods throughout Russia.

In addition to these species of *Boletus*, a number of other Hymenomycetes are common in Birch woods and many had already been cited as mycorrhiza formers. Woronin (1885 *a*) had named *B. scaber* and *B. edulis*; Frank (1892), *Amanita muscaria*; McDougall (1914), *Boletus* sp. and *Cortinarius* sp.; Peyronel, *Boletus scaber, Amanita muscaria, Lactarius necator* and *Scleroderma vulgare*. On account of their constant association with the respective trees, Smotlacha (1911) believed that *B. rufus* and *B. scaber* formed mycorrhiza in Aspen and Birch respectively.

Melin's researches were carried out by means of pure culture seedlings as described for Pine and Spruce, the ability to form synthetic mycorrhiza being tested and confirmed by inoculation from cultures isolated by appropriate methods from the sporophores of suspected Hymenomycetes.

Using similar methods, it was shown that many of the Hymenomycetes common in Birch woods are mycorrhiza formers. Decisive experimental proof has now been supplied by Melin for *Boletus scaber* and *B. rufus*, both of which formed mycorrhiza in Birch and Aspen, also for *Tricholoma flavobrunneum* and *Amanita muscaria* which formed mycorrhiza in Birch.

From previous work on Conifers it is regarded as probable that a number of other genera, notably species of *Lactarius, Cortinarius* and *Russula* are likewise responsible for mycorrhiza formation in Aspen and Birch. Moreover, although *Boletus edulis* did not form synthetic mycorrhiza under experimental conditions and appears only sporadically in Birch woods, it is probably a mycorrhiza former of the second order. *Boletus luteus* and *B. badius*, so characteristic of coniferous woods, were not observed to form mycorrhiza either in Birch or Aspen.

As was the case with Pine and Spruce, the experimental seedlings lacking infection grew freely, and showed similar modifications of the root system. It follows, therefore, that there is no obligate relation with the developmental seedling stages of these trees as in Orchids and Heaths.

Certain earlier observers, e.g. Brefeld (1908) and Duggar (1905) had noted the difficulty of cultivating the humus-dwelling Hymenomycetes on artificial media; the genera *Cortinarius, Lactarius, Amanita, Russula* and *Boletus* have all been named in this connection. These records, coupled with his own observations, led Melin to the view that such refractory species are possibly obligate mycorrhizal symbionts, incapable of growth, or growing with difficulty, in the absence of certain nutritive materials ordinarily obtained from the roots of their host trees.

The list of Hymenomycetes experimentally identified as mycorrhiza formers in Pine and Spruce has since been extended to include *Amanita muscaria*, *Cortinarius muscosus*, *Lactarius deliciosus* and *Russula fragilis* on Pine, and *Amanita muscaria*, *Cortinarius balteatus*, and *Lactarius deliciosus* in Spruce (Melin, 1925). No proof has yet been obtained as to the identity of the forms known respectively as *Mycelium radicis silvestris* β, γ, δ, or of *Mycelium r. abietis*.

Fresh light has been thrown upon the physiological aspects of mycorrhiza in forest trees in an experimental research designed by Melin (1925 *c*) to elucidate the physiology of nutrition in both symbionts and test the nature of their mutual relations. The tentative conclusions reached by the author himself are clear from his earlier papers. Thus:—"Der anatomische Bau des beschriebenen Mykorrhizatypus spricht also für einen gegenseitigen Stoffwechsel zwischen den beiden Konstituenten, d.h. dafür, dass wir es mit einer wirklichen mutualistischen Symbiose zu tun haben" (Melin, 1923 *a*, p. 94), and again, in relation to the mycorrhiza of Birch and Aspen:—"Die Birken- und Espenmykorrhiza ist kein parasitisches Gebilde, wie es z.B. McDougall meint, sondern Pilz und Wurzel leben in mutualistischer Symbiose miteinander" (Melin, 1923 *b*, p. 517).

These opinions survived the test of experiment. Valuable results were obtained from an experimental study of the root fungi of Pine and Spruce in pure culture, and the conclusions based upon them constitute an important contribution towards the solution of a much-discussed and highly controversial part of the mycorrhiza problem. Two aspects of this work demand attention: one, the nutritive reactions shown by the symbionts in pure culture with special reference to the assimilation of nitrogen; the other, the application of the experimental results to conditions found in nature.

As compared with indifferent soil species, the growth of the mycorrhizal fungi of Pine and Spruce is variable and often very slow. Some species, e.g. the *Boleti* associated with these two trees, are relatively vigorous, others, e.g. *M. radicis silvestris* β, *M. r. silvestris* γ, and *M. r. abietis*, grow extremely slowly, others again, e.g. *Russula fragilis* and *Amanita muscaria*, are maintained in culture with difficulty. Some species form sporophores in artificial media, others do not. Many show loss of vigour due to cultural conditions, and the physiological change thus suffered is maintained when the mycelium is subcultured to a favourable medium. Such inconsistencies of behaviour can be explained by the assumption that many of these fungi are obligate symbionts in greater or less degree, incapable of normal growth in the absence of their respective host trees. If this assumption be correct, the existence of such forms obviously limits the field of experimental enquiry.

Moreover, the mycorrhizal forms have been shown to be extremely sensitive to the H-ion concentration of the substratum. Whereas other soil fungi, e.g. *Rhizoctonia silvestris* and *Mycelium radicis atrovirens*, grow indifferently on acid and neutral media over a wide range of *p*H values, the true root fungi, with few exceptions, prefer acid substrata, grow badly on those with a neutral reaction and are incapable of growth at *p*H values on the alkaline side of neutrality. Optimum conditions for the fungi of Pine and Fir are provided by *p*H values between 4·0 and 5·0. The correctness of these experimental values receives support from Hesselman's (1917) field observations recording a *p*H value of about 4·0 in the humus layer of coniferous woods throughout northern and middle Europe. In Finland likewise, Brenner (1924) reported values from 3·5 to 4·8 for similar soils. Isolated observations on the H-ion concentration in root cells, e.g. those of Arrhenius (1922) recording a *p*H value of 4·5 for the root cells of Wheat have a possible significance in the same connection.

The reaction of the root fungi to small amounts of phosphatids is of particular interest in view of Hansteen-Cranner's observation that, under certain conditions, roots of the higher plants excrete phosphatids, not only when immersed in distilled water but also in nutrient salt solutions and fertile soil. (Hansteen-Cranner, 1922.) Furthermore, it is known that the growth of many species of bacteria is stimulated by root excretions (Wilson, 1921), (Barthel, 1923), and this fact, combined with his own observation that contact with roots of the appropriate host trees perceptibly stimulated the growth of mycorrhizal fungi, led Melin to examine the reaction of the latter to substances excreted from seeds and seedlings of Pine and Spruce[1].

The results obtained from cultures for which different nutrients and combinations of nutrients were used, indicated that the diffusible substances caused a marked stimulation of growth. For example, after 40 days in contact with seeds or seedlings, *Boletus variegatus* gave a dry weight value sixteen times as great, and *M. radicis*

[1] It had been stated previously (Hansteen-Cranner, 1922) that living seeds and seedlings of these trees give off appreciable amounts of (water soluble) phosphatids at 20° C.

abietis one fifty-six times as great as when grown alone. Control cultures in which dead seeds were substituted showed no effect of this kind. Similar results were obtained with ordinary soil fungi. Assuming that the concentration of phosphatids is proportional to the number of seeds used, the experiments indicate that an optimal concentration is quickly reached, after which no increase in growth can be observed. It was concluded by Melin that the effect is similar in kind to that brought about in *Penicillium* by the addition of an extract of Yeast (Lepeschin, 1924).

Experiments to test the reaction of the fungi of Pine to various nitrogenous food materials proved that none of the forms experimented with could utilise atmospheric nitrogen; salts of ammonia, urea, and nucleic acid were all used as sources of nitrogen, while individual fungi could equally well utilise peptone, asparagin and a number of organic compounds. Assuming enzyme activity in the mycelium to be comparable with that found in the sporophore, Melin's conclusion that mycorrhizal fungi in nature can utilise different organic compounds or groups of such compounds in humus soils may be safely accepted.

Variation in the nature of the carbon compounds supplied to cultures showed that growth was satisfactory only when glucose was added, thus providing confirmation of the popular view that mycorrhizal fungi obtain carbon compounds from the root cells. Direct observations on humus were difficult owing to the toxicity produced by heat sterilisation. Chemical methods of sterilisation were not regarded as satisfactory, but humus extracts freed from microorganisms by filtering gave weak growth, the addition of glucose producing vigorous development as in a favourable soil.

The reaction of uninfected seedlings of Pine and Spruce to various nutrients has also been experimentally tested. As might be expected, no evidence was found that seedlings in pure culture could fix atmospheric nitrogen. Inorganic compounds of nitrogen, e.g. potassium nitrate and ammonium chloride, proved to be favourable sources of nitrogen; simple organic compounds, e.g. asparagin, were also readily utilised, but more complex bodies like peptone and nucleic acid were assimilated with difficulty, and seedlings supplied with these substances soon showed symptoms of nitrogen starvation (Fig. 56).

From the experimental results yielded by seedlings and root fungi grown separately under pure culture conditions it was unlikely that synthetic cultures would exhibit any capacity to utilise atmospheric nitrogen. In view of its great importance this possibility was carefully tested and gave negative results, the small increase in nitrogen content in seedlings grown in substrata lacking combined nitrogen being regarded as due to impurities in the air of the laboratory. Melin has found no evidence of nitrogen fixation on the part of any coniferous tree and the claim put forward by Müller in respect to *Pinus montana* has received no experimental support.

The root development observed in experimental seedlings varied directly with the nature of the nitrogen supply, e.g. roots supplied with nucleic acid were from four to six times as long as those supplied with potassium nitrate, ammonium chloride or asparagin.

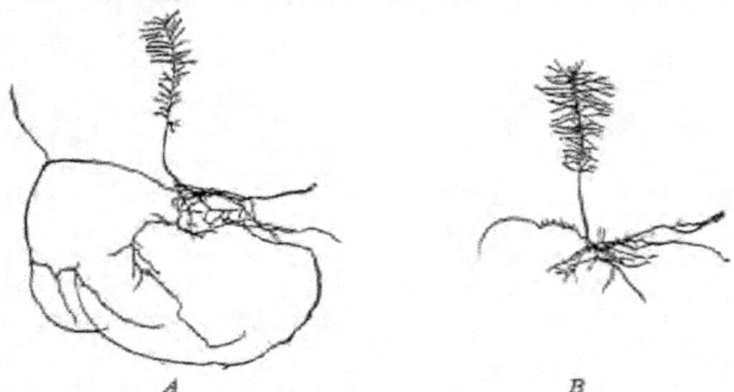

Fig. 56. Seedlings of *Picea abies*, three years old, in pure culture, with nucleic acid (Nukleinsäure) supplied as a source of nitrogen. *A*, seedling without fungus infection; *B*, seedling infected with the mycorrhizal fungus, *Mycelium radicis abietis*. About ½ natural size. (From Melin, 1925.)

The experiments made by Melin to test the reaction of infected as compared with uninfected seedlings towards various compounds of nitrogen are of special interest and importance in view of the historic controversy as to the nature of nutrition in tree mycorrhiza and its relation to the soil humus. The comparative growth of synthetic cultures as compared with that of pure culture seedlings supplied in each case with ammonium chloride, peptone or nucleic acid was measured by means of dry weight estimations. In general, the figures obtained and the condition of the cultures justify Melin's conclusion that the mycorrhizal fungi facilitate the intake of ammonium salts and other organic compounds by the tree, although it is not clear that such intake may not be carried out more rapidly under pure culture conditions by roots free from fungus infection.

By supplying more complex compounds, e.g. peptone or nucleic acid, evidence was obtained that these could be assimilated more readily by mycorrhiza than by uninfected roots, thus permitting the conclusion that mycorrhizas are organs directly related to the nutrition of the tree and having a special importance in relation to the assimilation of organic nitrogen. For example, for a number of three-year-old Pine seedlings, the average percentage nitrogen content was 2·86 in seedlings with mycorrhiza, as compared with 2·64 in those lacking infection. The prevalence and robust development of mycorrhiza in raw humus soils thus becomes extremely significant (Fig. 56).

In order to test the Stahl hypothesis in respect to inorganic salts, experiments were carried out with different concentrations and combinations of salts, final conclusions being based on a comparative study of the ash content of 50 plants. The values obtained (Melin, 1925, Table 43) pointed to the conclusion that inorganic mineral salts can be absorbed with equal facility by infected and uninfected roots in pure culture. On the whole the experimental results were in agreement with the Stahl hypothesis in respect to mineral salts, although the benefit derived by the trees was regarded by Melin as quite secondary to that associated with nitrogen assimilation.

With regard to the maintenance of a "balance of power" between the mycorrhizal fungi and their hosts, it is clear from the pure culture experiments described that the development of normal mycorrhiza is bound up with, and directly depends upon, the physiological state existing in each of the two symbionts, the health and vigour of the seedlings on the one hand, and the condition of the mycelium on the other hand, each playing an important rôle. The ectotrophic and endotrophic types of structure may be regarded as independent reaction products of the joint activities of mycelium and roots. Given an unhealthy condition of the latter and mycelium of high "virulence," the relation may degenerate into one-sided parasitism with the fungus in command of the situation. Assuming healthy root condition, the development of normal mycorrhiza depends upon those physiological qualities in the mycelium which constitute "virulence." In respect to this, Melin has proposed a classification of the root fungi studied into three classes described as *active, less active,* and *inactive* respectively. In relation to a particular host, members of the first class form typical mycorrhiza more or less easily under normal conditions; they are mycorrhiza builders of the first order. The fungi included in the second class produce mycorrhiza slowly and with difficulty; they are mycorrhiza builders of the second order. Those belonging to the third group are unable to form either ectotrophic or endotrophic mycorrhiza. In each class are included forms of slight, medium, and relatively high "virulence," as evidenced by their behaviour when brought into contact with seedling roots, their reaction in each case being determined both by the inherent specific qualities of the mycelium and by fluctuating changes in the environment. There is at present no full explanation of the observed fact that one and the same fungus may produce *intracellular* infection, and, under other conditions, give rise to an *intercellular* net without penetration of the cells.

Melin has correlated the failure to form mycorrhiza of the ectotrophic and endotrophic types in artificial cultures with decreased "virulence" on the part of the constituent root fungi. Seeking a cause for the latter condition, he determined experimentally that the true mycorrhiza formers grew feebly in substrata of unsuitable H-ion concentration, and finds in this fact an explanation both of his own experimental observations and of the distribution of coniferous mycorrhizas in nature. In view of the evidence offered, there can be little doubt that the reaction of the substratum is of great importance in determining the structural character of the mycorrhizal association. (Melin, 1924 *b.*) These observations have rendered it possible to attach a more precise physiological meaning to the varying degrees of "virulence" postulated by Melin. Exactly in what manner the biological relation is affected by the different types of structure observed in coniferous mycorrhizas is at present unknown.

No evidence has been obtained for the existence of different strains or *genotypes* of the same fungus species in association with different trees. For example, under cultural conditions one and the same strain of *Amanita muscaria* formed mycorrhiza in Pine, Spruce, Larch and Birch; one and the same strain of *M. r. silvestris* formed it in Pine, Spruce and Birch.

The application of these results to natural conditions is of great practical interest. In view of the experimental observations, factors likely to be effective are:—(1) the H-ion concentration of the soil, (2) the condition of the organic constituents of the soil, (3) the presence in soil of substances which inhibit growth of the root fungi. With regard to the first of the factors, Hesselman (1917) and others have reported *p*H values of 4·0 to 5·0 in coniferous humus throughout Sweden, and in such acid soils (*Rohhumusböden*) mycorrhiza is freely produced. In deciduous woodlands Hesselman (1917) and Arrhenius

(1920) found pH values between 6·0 and 7·0. In these neutral and slightly acid soils (*Mullböden*) Pine and Spruce form mycorrhiza only sporadically, although the absorbing roots frequently develop pseudomycorrhiza as a result of invasion by relatively inactive forms or by indifferent soil fungi. The variable development of mycorrhiza observed in different types of acid soils may be related directly to the second set of factors named above, namely, the presence of organic compounds of nitrogen acceptable to the mycorrhizal fungi, the observed concentration of mycorrhiza in the upper layers of humus being possibly due to the same cause. It is not without significance that the true root fungi show greater sensitiveness in this respect than do *Rhizoctonia silvestris, M. r. atrovirens* and casual soil fungi generally.

With regard to the existence in natural woodland soils of substances deleterious to the growth of mycorrhizal fungi there is at present no information. Under experimental conditions these forms are extremely sensitive to the accumulated by-products of growth in the substratum, and mycelium from such poisoned cultures does not build mycorrhiza. Viewed in the light of modern work upon soil conditions, the experimental results point to the conclusion that Conifers are not strictly autotrophic in respect to their nitrogen metabolism when growing on humus soils. In Melin's words:—"Die Mykorrhizen sind auf Rohhumusböden sehr günstige stickstoffvermittelnde Organe, da die Pilzsymbiont ebenso wie die anderen Bodenpilze leicht Ammoniak und organische N-Verbindungen assimilieren können." If this conclusion be correct, trees possessing typical mycorrhiza are remarkably well equipped for competition with members of the soil microflora under the conditions stated.

It is more doubtful whether the mycotrophic habit is indispensable to the healthy growth of Conifers on neutral humus soils. On the other hand, well marked differences are known to exist in such soils in respect to the rate at which nitrification takes place, and the conditions responsible for such differences may be causally related to an irregular development of mycorrhiza.

Passing now to the effect of mycorrhiza formation upon the constituent fungi, it may be noted that the stimulating effect produced by contact with the host tree appears to be related definitely to the excretion of phosphatids. It has been suggested by Melin that certain of the mycorrhizal fungi may be unable to complete their development, e.g. form fruit bodies, lacking this stimulus. This assumption requires experimental verification. If correct, it provides a satisfactory explanation of the observed disappearance of Hymenomycetes from cleared woodlands. In the evolutionary sense, the fungi concerned in mycorrhiza formation may probably be regarded as members of the saprophytic soil flora which have become adapted to symbiotic existence, although, in certain cases, e.g. *Tricholoma* sp., they remain relatively independent of it. Assuming the correctness of Melin's general conclusion:—i.e. "dass die Mykorrhizen für die Pflänzchen und Bäume auf Rohhumusböden und verwandten Bodentypen eine vitale Bedeutung besitzen," the application of the new knowledge gained from experiment becomes a matter of great practical interest, and opens up a fascinating new field of research in relation to forestry.

His work has confirmed the view certain to be reached by every modern student of mycorrhiza; namely, that there is no sharp distinction of the kind postulated by Frank between the ectotrophic and endotrophic types of structure. The formation of an external sheath and the intercellular development known historically as the "Hartig net" do not preclude the existence of an intracellular distribution of the same mycelium. Recent experiments have shown that the type of mycorrhiza formed is controlled, in the case of certain conifers, by the physiological condition of the endophyte and by the rooting condition of the host. It is, therefore, a resultant of these conditions and must be regarded as an expression of the physiological activities of the two symbionts at the time of observation. The factors that influence these activities are still relatively little known; their study offers a profitable field of enquiry and one which must be more fully explored if there is to be any real understanding of the nutritive relations between fungus and vascular plant in the case of trees.

Certain recent researches on the decomposition of woodland humus point to conclusions significant in this connection.

In humus soils the liberation of plant food from the organic constituents is known to depend upon chemical changes brought about by the nutritive processes of members of the soil microflora and microfauna. Since these processes involve complex chemical changes, the subject is extremely intricate; it is also of great practical importance in relation to forestry. Although not necessarily or immediately concerned with mycorrhiza, it is closely bound up with the activities of soil fungi, and hence comes into touch with the ecological and physiological aspects of root infection.

The deterioration of forest soils resulting from an abnormal accumulation of raw humus is stated by Falck (1923) to be directly related to the absence of certain members of the fungus flora. In fertile forest soils, certain members of the Higher Fungi, especially Hymenomycetes and other Basidiomycetes, with members of the Mucoraceae, are believed to be primarily responsible for the complete decomposition of fragments of wood, leaves and other plant residues. Falck has investigated and described the chemical changes involved in this process, to which he has given the name "Mykokrinie." His researches show likewise that the normal process of decomposition by fungi is often interrupted by the activities of insect larvae, which devour both the mycelium and the detritus on which it is growing, converting it eventually to a dark-coloured crumbling humus mass—a process described under the name of "Anthrakrinie." In humus soils undergoing this kind of decomposition four layers may be distinguished:—

(1) An upper layer of unchanged plant detritus;

(2) A region in which mycelial activity is conspicuous, especially in spring and early summer;

(3) The humus layer proper, composed below chiefly of dark-coloured insect excreta and passing upwards into partially decomposed vegetable detritus;

(4) The uppermost layer of mineral soil containing the soluble material from the humus layer above leached out by rain, and traversed by a dense network of tree roots.

The deterioration of forest soils, due to an abnormal accumulation of raw humus and consequent deficiency of available nutrients, is believed to be due mainly to the absence of the ordinary woodland type of fungus flora. Failing the activities of suitable fungi, leaves and other residues collect upon the surface and undergo chemical changes of an entirely different kind, resulting in the conversion of the raw humus to "dry peat" (Trockentorf) in which the carbon, nitrogen, phosphorus, and potassium are locked up and rendered permanently inaccessible to the trees. This formation of peat or "Vertorfung" is named "Anthragenie," and can be related by transitional types of change to the complete decomposition taking place in *mycocriny* and *anthracriny*.

The factors promoting the latter are believed to be the following: high atmospheric humidity associated with high soil temperature; a sufficiency of lime; an abundant and suitable fungus flora; the favourable character of the detritus.

PLATE VI

The relatively satisfactory development of many kinds of trees on dry peat or other humus soils in which no progressive decomposition is taking place is explained by an appeal to the mycotrophic habit, i.e. by the assimilation of organic nutriment from humus-forming detritus with the help of mycorrhizal fungi.

The benefits accruing to forest trees from symbiotic relations with mycorrhizal fungi—excluding nodule-forming associations, in which nitrogen fixation occurs—consist, in Falck's opinion, solely in the presentation of the requisite foodstuffs in the form of organic solutions.

The advantages of *mycocriny*, *anthracriny* and the mycotrophic habit as compared with *anthrageny*, in relation to practical problems of forestry, are discussed by Falck, and he points out that trees with well-developed mycorrhiza are far better adapted to unfavourable climatic and soil conditions than are those devoid of fungus symbionts. Suggestions are made also for promoting the decomposition of humus by biological methods.

Melin has reported that soil Hyphomycetes do not form mycorrhiza in trees, although certain members of the group, e.g. species of *Rhizoctonia*, are known to invade roots more or less parasitically and to give rise to pseudomycorrhiza. Viewed in the light of Falck's observations on the distribution of fungi in forest soils, this con clusion is not without interest.

The ectotrophic habit is typically associated with trees and shrubs. For this reason, in the present chapter, attention has been focussed upon the mycorrhiza of trees, more especially upon those recent experimental researches from which alone can be gained any knowledge as to the real significance of the habit. A few isolated cases of this type of mycorrhiza are known in herbaceous plants, notably the classical one of *Monotropa* and its allies. Most of these have received mention elsewhere in the present work; in none of them has the biological character of the relation yet been illuminated by experimental enquiry.

EXPLANATION OF PLATE VI

Fig. 50. "Knollenmykorrhiza" and "Gabelmykorrhiza" borne on the same main root of Pine (*P. sylvestris*). (From Melin, 1923.)

Fig. 51. Mycorrhiza of Beech in superficial deposits of fallen leaves below the trees.

Fig. 52. "Fairy ring" of *Clitocybe* sp. under Larch, Co. Wicklow, Ireland.

CHAPTER IX

FUNGUS INFECTION IN BRYOPHYTA

"Mycorrhiza" in Bryophyta: fungus infection in Liverworts; Nemêc; Stahl; Gallaud; Cavers; Ridler; Bernard; Magrou; the characters of the fungi concerned—Infection in Mosses—Mycorrhiza in the Pteridophyta; Equisetales—Lycopodiales—The gametophyte of *Lycopodium*: historical; Treub; Bruchmann; Lang—*Psilotum*: Shibata—The sporophyte of *Lycopodium*—Ophioglossales—Filicales; Marattiaceae—Filices; fungus infection in Osmundaceae, Gleicheniaceae and Cyatheaceae—The absence of mycorrhiza in Polypodiaceae—New records of its occurrence in *Aspidium* and *Pteridium*.

Bryophyta.

THE regular occurrence of mycelium in the tissues of members of Bryophyta has long been known. There appears to be little doubt that the relation is frequently of similar biological nature to that in mycorrhiza and the use of this term to describe it has been sanctioned by usage, although it involves a very loose use of words and appears to the writer to serve no useful purpose.

The history of research on the subject reproduces in miniature that on mycorrhiza proper, i.e. it shows an earlier period of isolated observations and speculations leading gradually to the development of a more critical type of experimental investigation. Gottsche and Schleiden (1843) had noticed spiral threads in the cells of *Priessia*, the fungal character of which was afterwards recognised by Schacht (1854) and by Gottsche himself in 1858. Another early record was made by Leitgeb (1874–1881) who described fungal infection in the young sporogonium of *Ptilidium ciliare*. Janse (1897) observed it in *Zoopsis*, a tropical member of Jungermanniaceae native in Java.

Nemêc (1899) recorded the presence of mycelium as a regular occurrence in the leafy members of Jungermanniaceae, of which all the species examined, with the exception of *Jungermannia tridentata*, were found to be infected. Discussing the incidence of infection, he concluded that it was dependent on habitat, and mentioned the case of *Lepidozia reptans*, a species showing typical infection in humus-rich soil in shady situations but found growing on clay soils entirely free from mycelium. Describing the infection of the rhizoids and penetration of the neighbouring stem tissues in *Calypogeia trichomanes*, Nemêc summed up in favour of the existence of a reciprocal relation as follows:—

Man könnte die zäpchenformigen Fortsätze vielleicht als Haustorien deuten, welche der Pilz in die Zellen der Wirthspflanze einsendet, um Nährstoffe aus ihnen saugen zu können. Es ist jedoch anderseits ebenso möglich, dass die Wirthspflanze die Pilzhyphen zu derartigen Gebilden reizt, um an einer grossen Oberfläche und bei inniger Berührung möglichst leicht Stoffe entnehmen zu können.

The same author recorded mycelium in the rhizoids of *Kantia trichomanes*, *Lepidozia reptans* and *Lophozia bicrenata*. In *Kantia*, the mycelium was believed to be that of *Mollisia jungermanniae*, an Ascomycete with small bluish green apothecia, often found on the thallus. Nemêc observed that infected cells in *Kantia* retained their contents, the nuclei placing themselves in close proximity to the invading hyphae.

In members of Jungermanniaceae infection is usually conspicuous in the rhizoids, from the proximal ends of which interweaving branches form a pseudo-parenchymatous tissue whence outgrowing hyphae penetrate the stem. While Nemêc noted that fungal infection was general in this group of Liverworts, he believed it to be absent from members of Marchantiaceae, even when growing in close proximity to infected species. Stahl (1900) attempted to correlate the incidence of infection in the Hepaticae with his theory of nutrition in mycotrophic plants. He identified the members of Jungermanniaceae and Marchantiaceae respectively as sugar-containing and starch-containing forms, and proposed to investigate the possibility of a parallelism between fungus-infection and the water-balance among Liverworts. Unfortunately for Stahl's hypothesis, the supposed differential behaviour of members of the two groups in respect to infection was not confirmed, and evidence was soon forthcoming that "mycorrhiza" formation was as frequent a phenomenon in Marchantiaceae as in Jungermanniaceae. Indeed, the presence of mycelium in the rhizoids of Marchantia and in those of the common Liverwort *Lunularia vulgaris* had already been noted by Kny (1879) who observed also that infection extended to the ventral parts of the thalli when growing on a substratum rich in humus.

Beauverie (1902) noted the presence of mycelium in the thallus of *Fegatella*, another member of Marchantiaceae, and believed the fungus present to be related to *Fusarium*. Golenkin (1902) added fresh records for the Marchantiaceae, describing endotrophic infection in *Marchantia palmata*, *M. palmacea*, *Priessia commuta*,

Targionia hypophylla, *Plagiochasma elongatum* and *Fegatella conica*. This author compared the endophytes with those of *Neottia* and *Lycopodium*, and suggested that the infected tissue functioned for water-storage. Against this view Cavers (1903) cited his own observations on *Fegatella* and *Monoclea*, both typically hygrophilous forms. Moreover, the thallus of the former contains well-developed mucilage tissue for the storage of water.

To the genera of Jungermanniaceae recorded by Nemêc as mycorrhiza-formers, Cavers added the names of *Cephalozia bicuspidata*, *Scapania nemorosa*, *Diplophyllum albicans*, *Plagiochila asplenioides*, *Bazzania trilobata* and *Porella platyphylla*, in all cases the degree of infection increasing with the amount of humus in the substratum. *Monoclea forsteri*, a New Zealand species, was also found to be heavily infected.

In every plant examined, vertical sections of the thallus showed a sharply defined mycorrhizal zone, consisting of from two to four layers of cells densely filled with branching fungal hyphae. This zone is confined to the thicker median portion of the thallus and extends to within a short distance of the growing point.

Gallaud (1905) held that the endophytes of Liverworts were so specialised in character as to form a distinct group, and described the intracellular distribution of mycelium and the development of vesicles and sporangioles. He also put on record the irregularity of infection observed in *Pellia*:—"Dans une même station de *Pellia epiphylla*, tous les thalles ne sont pas infestés, et ceux qui le sont le sont inégalement, sans qu'il soit possible, d'ailleurs, d'établir entre eux d'autre différence qu'on puisse nettement attribuer à l'infection." In general, he concluded that fungus infection was widely distributed in Hepatics, alike in Marchantiaceae and in Jungermanniaceae, and he believed that it was limited to the gametophyte generation. It was absent from or very rare in the region of the thallus near the sporogonium.

Humphrey (1906), investigating the development of *Fossombronia longiseta*, a member of a genus intermediate in position between the thallose Jungermanniaceae and the higher foliose types, noted the occurrence of tuberous swellings on the stems. Detailed examination of the tissues in these stem enlargements revealed in every case the presence of mycelium. The hyphae were restricted almost entirely to the stem and rhizoids, the latter sometimes exhibiting lateral swellings suggesting the formation of short branches. The existence of a causal relation between infection and tuberisation was not investigated, and there is as yet no information as to the character of the fungus in *Fossombronia* or its biologic significance to the thallus.

In view of the formation of tubers by other Liverworts—they have been recorded in *Geothallus tuberosus, Riccia perennis, R. bulbifera, Anthoceros tuberosus, A. phymatodes, Petalophyllum preissia,* while Goebel described and figured them in *Fossombronia tuberosa,* and Howe had noted their appearance in *F. longiseta*—more information respecting the biology of infection will be awaited with interest.

As might be expected from its nature, the evidence is inconclusive concerning the physiology of infection in Liverworts. The amount of experimental data is scanty. Cavers grew spore cultures on sterilised soil and believed that development was more vigorous when mycelium was present. In control cultures he noted also that infection took place more readily from soil containing abundant humus. In general, observers have tended to regard infection as a symbiotic relation enabling the host to assume a partially saprophytic method of nutrition. In respect to the distribution of mycelium and restriction of infection observed in Liverworts, it is worth recalling the fact that Czapek (1889) had recorded the presence of an antiseptic substance, *sphagnol,* in the cell-walls of many members of Hepaticae.

There is a long gap in the records after 1906 and subsequent contributions to the knowledge of fungus infection in Liverworts fall well within the modern period. Ridler (1922) confirmed Gallaud's observations on *Pellia* in respect to the irregular distribution of infection and believed it to vary with the age of the thallus rather than with the habitat. In fronds showing the typical condition, mycelium was found to be localised to the rhizoids and the thicker part of the thallus forming the midrib—"the upper two or three layers, including the upper epidermis, remains free from hyphae."

Gallaud had described the morphology of the endophyte in *Pellia*; main trunk hyphae of large diameter bearing vesicles and sporangioles; while the presence of arbuscules, although not recorded, is implied in that of the latter organs. Ridler (1923) corroborated these observations and also described and figured the arbuscules composed of "very fine and profusely branching threads" which completely filled some cells of the thallus with a network of mycelium. The relation of these structures to the so-called sporangioles was well shown in the figures supplied by this author.

Nicolas (1924) has recently added a new record for *Lunularia,* describing regular endotrophic infection in certain thalli bearing antheridia, although mycelium was absent from neighbouring sterile plants. As is well known, the formation of sexual organs by this common Liverwort is an exceptional occurrence, but Nicolas' attempt to establish a causal connection between the two phenomena was unsuccessful inasmuch as he subsequently discovered antheridia-bearing plants free from mycelium.

Certain recent observations on an Indian species of *Marchantia* are of interest. It was noted by Chaudhuri and Rajarum (1925) that vigorous plants of *Marchantia nepalensis*, collected in the neighbourhood of Lahore, invariably showed fungus infection of the thalli. The mycelium was confined to the gametophyte and restricted to a zone of tissue below the air canals.

Isolation experiments yielded a non-sporing mycelium which readily invaded fresh gemmae giving rise to the characteristic infection. The endophyte was cultivated on various artificial media and throve on those with a reaction of *p*H 6·6 to 7·0. Deprived of maltose, the mycelium ceased to grow, whereas withdrawal of asparagin produced little effect. It was inferred that the organism depended upon its host for a supply of carbohydrates. Comparative cultures of infected and uninfected *Marchantia* thalli on filter paper or sterilised soil showed that those lacking mycelium developed normally, but died off without producing sporophytes. It was concluded by the author that the condition in this species of *Marchantia* is one of reciprocal symbiosis, infection by a specific fungus being essential to full development of the green plant.

Among the more important papers published by Noël Bernard in the course of his work on Orchid mycorrhiza was one dealing with the evolution of plants in relation to symbiosis—*L'Évolution dans la Symbiose* (Bernard, 1909*a*), in which the author elaborated his thesis that the advanced degree of adaptation shown by certain members of the Bryophyta to symbiotic relations with fungi might provide a clue pointing to the origin of vascular plants from members of this phylum.

The hypothesis was based mainly on Bernard's comparative observations on the relation of tuberisation to fungus infection, together with Treub's researches on the comparative embryology of the Lycopodiaceae, as a result of which, the latter observer had been led to the view that living Vascular Cryptogams were descended from plants resembling certain Bryophytes—"quant à l'essentiel aux Muscinées actuelles (plus particulièrement aux Hépatiques)" (Treub, 1890).

With the intention of providing experimental evidence bearing on this evolutionary theory, a study of the relations in *Pellia* was undertaken by Magrou (1925). The account contributed recently by this observer provides a consistent explanation of the irregularities of infection observed in this common Liverwort and his observations on the distribution of mycelium in thalli of *Pellia* growing under natural conditions are in agreement with those on the same subject made earlier by Bernard and recorded for the first time in this paper. In particular, Bernard had noted that strongly growing fertile fronds were immune to infection up to the period of spore dispersal, becoming susceptible to infection subsequently. Magrou has confirmed this observation and recognised the presence in early spring of at least three generations of superimposed branches in any actively growing colony of *Pellia*, viz. old brown thalli with scars of sporogonia of the preceding year; younger thalli, dark green in colour, that have just produced sporogonia; and still younger fronds of bright green colour bearing antheridia and archegonia. The oldest fronds showed profuse infection by the characteristic endophyte, with abundant evidence of the effective control exerted by the host, by the gradual conversion of arbuscules to sporangioles within the cells of the thallus.

In the younger generation of fronds, infection was found to be less pronounced; in every case, a zone of tissue about the base of the young sporogonium exhibited immunity to infection by the rapid destruction, not only of haustorial branch systems, but of the main hyphal branches. This protective zone included the rhizoids developing on the ventral side of the same region. In the youngest generation of branches bearing sexual organs, the immunity secured in this way was even more conspicuous. Except in those cells most remote from archegonia, the mycelium showed complete degeneration, following so rapidly upon penetration as to cause almost complete inhibition of arbuscule formation. This immunity to infection was limited to tissues neighbouring living archegonia and actively developing sporophytes, and came to an end when these organs completed their development or became disorganised; following dehiscence of the capsule, mycelium rapidly invaded the surrounding tissues of the thallus. The endophyte was of the characteristic mycorrhizal type, producing vesicles and arbuscules (Fig. 57 *A*, *B*). Sowings of spores upon infected soil provided young gametophytes subjected to infection at an early stage of development. Observations on infected cells showed that the mycelium caused destruction of the chloroplasts, leading to the browning of tissue commonly recorded. It resulted also in marked arrest of growth, transferred ultimately to branches developing from islets of meristematic tissue, thus producing a characteristic tufted habit of growth.

To estimate correctly the exact rôle of the endophyte involves, in this case as in others, the separation and observation of each member of the association in pure culture, and their synthesis under controlled conditions. Inasmuch as the endophyte of *Pellia* has hitherto resisted isolation, it has been possible to draw only limited

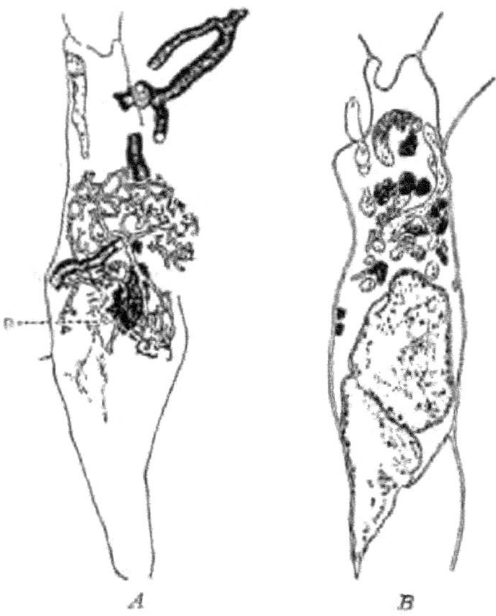

A *B*

Fig. 57. *A*, *Pellia epiphylla:* arbuscule in a cell of the thallus. *n*, cell-nucleus. Original × 695. Reduced about ⅓. *B*, *Pellia epiphylla:* cell of the thallus enclosing two multinucleate vesicles. Original × 695. Reduced about ⅓. (From Magrou, *Ann. d. Sc. Nat.* 1925.)

deductions from experimental cultures. The success of aseptic spore cultures, whether upon sterilised soil or upon artificial media, was found to depend upon the existence in the substratum of a H-ion concentration corresponding to that of the soil upon which the parent plant had been growing in nature, viz. a *p*H value of about 4·85. Under such conditions, thalli in aseptic cultures showed rapid growth and uninterrupted apical development. By comparing plants so obtained with infected thalli, Magrou has inferred the existence of alternating phases of rest and differentiation determined by infection.

At the time of publication (one year from sowing), sexual organs had not been formed by the plants in these cultures[1].

It has been pointed out by Magrou that a causal relation may exist between the high degree of acidity found necessary in his aseptic cultures of *Pellia* and the absence of mycelium in the absorbing region of the thallus. It is clear that this condition is not invariably fulfilled in nature, e.g. his experimental plants were derived from members of a colony upon acid soil ($p\text{H} = 4{\cdot}85$), whereas those used by Ridler were found growing upon soils the reaction of which was approximately neutral ($p\text{H} = 6{\cdot}8$ to $7{\cdot}0$). If the endophytic mycelium plays any part in the regulation either of the osmotic equilibrium or of the H-ion concentration of the cell sap this discrepancy might be susceptible of explanation.

Recognition that the proximity of archegonia and young sporophytes constitutes a definite bar to the spread of mycelium in *Pellia* has helped to clear up confusion in the literature in respect to sporophyte infection in Liverworts. It had been stated by more than one of the earlier workers, including Gallaud, that fungal infection in Bryophyta is limited to the gametophyte. There then appeared a series of records of sporogonial infection—Leitgeb (1879) for *Ptilidium*, Cavers (1903) for *Lophocolea tridentata*, *Radula complanata*, *Cephalozia bicuspidata*, and *Plagiochila asplenoides*. Reviewing these observations in the light of his own researches, Cavers concluded that the relation in sporophyte infection was "simply that of parasitism."

More recently Ridler (1922), working on *Pellia*, after recording that—"Normal healthy sporophytes were repeatedly examined, and no sign of the fungus was discovered in foot, seta or capsule"—observed subsequently that mycelium was present in certain abnormal capsules, and isolated a species of *Phoma* from the infected tissues. At the time of the observation, the account given implied identity of this mycelium with that present in the tissue of the thallus. It was concluded from the condition of the infected sporogonial tissues that the relationship was parasitic on the part of the fungus:—"The fungus causes a disease, killing the tissues of the sporogonium, and in some cases rendering them abortive." Attempts to obtain experimental proof by inoculation into uninfected thalli

[1] Killian (1923–1924) has recorded the production of sporogonia by various members of the Jungermanniaceae in pure cultures. He attributed lack of success in the case of *Pellia* to the absence of the specific endophyte, and has undertaken further experiments with a view to securing further evidence on the subject.

were not attended by success, and it was admitted later that the confirmation of the identity of the two mycelia was lacking (Ridler, 1923). On general grounds there appears little reason to believe that they belong to fungi related in any way to one another. All the facts recorded and figured point to pathogenic infection of the sporophyte by a parasitic fungus, doubtless the species of *Phoma* isolated by Ridler. Viewed thus, the case falls into line with others recording sporophytic infection in Liverworts, and confirms rather than contradicts the theory of sporophytic immunity offered by Magrou.

As a transitional group from the thalloid to the foliose types, the Aneuraceae have a somewhat special interest. It was known that species of *Aneura*, e.g. *A. pinguis*, showed fungus infection of the rhizoids and neighbouring cells. Denis (1919) described a "biologic form" of *Aneura* sp. without chlorophyll although apparently healthy since normal sporogonia were produced on some of the fronds. These thalli showed profuse intracellular infection of the rhizoids and ventral tissues by an endophytic fungus which formed typical "pelotons" in the cells. The mycelium resembled that noted in *Aneura pinguis*, but the colourless thalli showed markedly heavier infection, held by Denis to be directly related to the absence of chlorophyll. No experimental data were obtained. An infection of similar type in a normal green species of *Aneura* is shown in the photograph reproduced in Fig. 58. In this case likewise the mycelium is similar in habit to that of the Orchid endophytes. It forms "pelotons" which are subject to digestion by the cells of the host. This kind of infection appears to be constant for *Aneura*, and is markedly different from that recorded in members of the Marchantiaceae and Jungermanniaceae, in which the endophytes belong to the familiar mycorrhizal type producing vesicles and arbuscules. In view of Peyronel's recent conclusions respecting the presence of two different endophytes in the mycorrhiza of Flowering Plants, this observation is noteworthy. Whether infection by both types of fungi can occur simultaneously in Liverworts is at present unknown. (Pl. VII, Fig. 58.)

In respect to fungus infection in the higher members of Bryophyta, endotrophic mycelium is well known to occur regularly in certain Moss genera, e.g. *Buxbaumia* and *Tetraplodon*. Peklo (1903) described it in the capsule of *Buxbaumia aphylla* and interpreted it as a case of "mycorrhizal association." Cooke (1889) and Britton (1911) recorded *Cladosporium epibryum* on various Mosses. Györffy (1911) noted the presence of *C. herbarum* on capsules of *Buxbaumia viridis*.

Schimper (1858) and subsequent workers had observed the presence of small "microspores" together with spores of normal size in capsules of *Sphagnum*; the former were regarded as "male spores" but were afterwards identified as the chlamydospores of a "Smut" fungus, *Tilletia sphagni* (Nawaschin, 1892).

In general, it may be concluded that there is practically no evidence at present for the existence of anything in the nature of a mutualistic relation with fungus mycelium in Mosses as a whole, while the claim for its existence in certain specialised genera, e.g. *Buxbaumia* and *Tetraplodon*, lacks experimental confirmation.

Pteridophyta.

In Vascular Cryptogams, fungus infection may involve either phase in the life history or affect both gametophyte and sporophyte generations. The word "mycorrhiza," to describe an association of fungus mycelium with the prothallus, is obviously a misnomer but, as in Bryophytes, its use has been sanctioned by custom. Since the biological relation is evidently similar and may even involve the same fungus in both generations, it is clear that any full account of the condition in Pteridophytes must include both phases.

Equisetales. It is curious that there is no record of mycorrhiza in the genus *Equisetum*, several species of which frequent soils rich in humus. In *Equisetum arvense* and similar forms, the deep-growing habit of the rhizomes may afford an explanation of its absence. The association of a reduced transpiring surface and other indications of xerophily with absence of fungus infection appeared to offer a direct challenge to the theory of infection put forward by Stahl, but an experimental enquiry by this observer permitted him to record the existence of a relatively rapid transpiration current in species of *Equisetum*.

The invasion of the prothallus of *Equisetum* by a species of *Pythium*, put on record by Sadebeck (1875), appears to have been merely an isolated case of ordinary parasitic attack.

Lycopodiales. In this group of Pteridophytes, most of the interest relating to fungus infection is centred in the gametophyte stage.

The prothallus of *Lycopodium* is still a botanical rarity, known to comparatively few botanists. This is not due to lack of interest in the genus, but to the two circumstances that the small, subterranean gametophytes are difficult to discover in nature and have only recently been collected in any abundance, while their cultivation from spores presents difficulties.

The history of attempted spore germination goes back to the end of the eighteenth century, when an English surgeon, John Lindsay (1794), reported successful germination of the spores of *Lycopodium cernuum*, but unfortunately left on record no details beyond the fact of "a vegetable growth taking place where they (the spores) were sown." Spring (1842) failed to secure germination and attributed his failure to the existence of male plants only! This was before Hofmeister (1851) had correctly predicted that spores of *Lycopodium* would produce—not leafy plants, as was generally assumed at the time—but prothalli bearing sexual organs. Hofmeister, unfortunately, failed to secure proof of this by means of spore germination. Some success was achieved by de Bary (1858) who germinated spores of *Lycopodium inundatum*, but was unable to maintain the young prothalli beyond a ten- or eleven-celled stage. Beck (1880) also germinated spores of this species with a similar result.

The first discovery of gametophytes in nature was made by Fankhouser (1873), who found four prothalli of *Lycopodium annotinum*, with sporelings attached. These, combined with the early stages of development previously observed by de Bary, provided a general idea of the structure of the gametophyte of *Lycopodium*. Large contributions to the knowledge of this still somewhat mysterious stage in the life history were then made by three or four botanists, of whom two, at least, carried out researches now classical in the history of the subject. Goebel (1887) found and described prothalli of *L. inundatum*, thus completing the story begun by de Bary nearly thirty years earlier.

In 1884 Treub (1884–1890) published the first of his elegant researches on tropical species of *Lycopodium*. In a series of papers entitled *Études sur les Locopodiacees* he figured and described the development and structure of the gametophytes and young sporophytes of a number of species as found in nature, and recorded also the results of a series of laboratory cultures of the spores. These accounts have become a part of standard botanical literature and need not be recapitulated here. The prothalli described included partially subterranean cylindrical forms with chlorophyll in the upper region (*L. salakense*), similar forms with leaf-like green expansions at the surface (*L. cernuum*), tuberous forms without chlorophyll, entirely subterranean in habit (*L. clavatum*), and the colourless branched saprophytic forms belonging to *L. phlegmaria* and related epiphytic species. Treub germinated the spores of *L. cernuum* in the laboratory on soil brought from stations carrying the plant in the field and, under these conditions, succeeded in growing the prothalli to a relatively advanced stage of development. They showed characteristic infection by an endophytic fungus, the mycelium occupying the cells of a median zone in the lower, colourless part of the prothalli.

Treub thus described infection in the prothallus of *L. cernuum*:—"Les cellules périphériques des tubercules renferment toutes les filaments d'un champignon endophyte appartenant probablement au genre *Pythium*." The mycelium was found in all the prothalli examined, and was apparently harmless to the cells (Treub, 1885).

To Treub the most striking feature of infection was the distribution of the hyphae, intracellular in the peripheral regions, and strictly intercellular in the more centrally situated tissues. Subsequently, he described the gametophytes of a number of other species—*L. salakense, L. phlegmaria, L. hippuris*, and *L. nummularifolia*. *L. salakense* has a green aerial gametophyte. When sown, the spores germinate at once, and, in this species only, Treub succeeded without difficulty in raising mature prothalli with antheridia and archegonia. There is no mention of fungus infection in these prothalli, and Treub regarded them as holophytic in nutrition, although they lacked the green leaf-like expansions which crown those of *L. cernuum*:—"En effet, les prothalles de *L. salakense* ne vivent certainement pas en saprophytes; ils sont distinctement verts, bien qu'ils ne contiennent pas autant de chlorophylle que ceux de *L. cernuum*" (Treub, 1887–1888).

The gametophyte of *L. phlegmaria* is a branched structure without chlorophyll, bearing gemmae or bulbils which grow directly into new prothalli. The distribution of the endophytic fungus is entirely intracellular in this species. Those of *L. carinatum, L. hippuris* and *L. nummulariforme* are of similar habit. Treub failed to germinate the spores of these saprophytic forms. In the case of *L. phlegmaria*, he put on record his conviction that the relation with the endophyte was not parasitic:—"le champignon abrité par le prothalle pourrait payer le service rendu en contribuant à la nourriture de son hôte. Il y aurait mutualisme."

While Treub's researches were in progress, Frank's papers on mycorrhiza appeared and his theory of beneficial symbiosis challenged attention. Treub, however, was hardly prepared to admit this relation for *Lycopodium* prothalli in general. In *L. cernuum* and *L. inundatum*, for example, he inclined to the view that the endophytes were harmless intruders—"de faire peu de mal" or "de ne pas tuer les prothalles." With regard to the identity of the endophytes, he had originally referred that of *L. cernuum* to the genus *Pythium*. In view of his own later observations and those of Wahrlich (1887) on the root fungi of Orchids, he did not press this suggestion, nor did he put forward any subsequent views on the subject.

Simultaneously with Treub's work on the tropical species, Bruchmann's (1885–1898) patient and protracted investigations on north European species of *Lycopodium* revealed the full life history of those much more difficult forms with subterranean prothalli. Bruchmann (1885) found and described the gametophyte of *L. annotinum*, and, thirteen years later, those of *L. clavatum*, *L. complanatum*, and *L. selago*. After repeated efforts, he succeeded in germinating spores of several species, a remarkable feature being the extreme slowness of germination. For example, spores of *L. selago* began to germinate 3–5 years from sowing, while the mature prothallus took 6–8 years to develop; those of *L. clavatum* and *L. annotinum* were even slower, requiring from 12–15 years to reach sexual maturity. The gametophytes of all these species are small subterranean tuberous structures, destitute of chlorophyll, and apparently saprophytic in nutrition.

Bruchmann recorded and figured mycelium in the rhizoids and adjoining tissues, and observed invasion of the young prothallus at the four- or five-celled stage by an endophytic fungus. He suggested that it might be possible to shorten the protracted germination period by supplying a suitable stimulant, but did not carry out any experiments on this aspect of the subject.

Almost simultaneously with Bruchmann, Lang (1899) discovered and described the gametophyte of *L. clavatum*. The prothalli of this species are very small tuberous bodies, entirely subterranean, and destitute of chlorophyll. In the main, the distribution of infection is similar to that of the other saprophytic forms—an outer layer of cells apparently free from mycelium, a peripheral zone showing profuse intracellular infection, and a central region, with hyphae confined to the intercellular space system, produced by separation of adjacent cells along the middle lamellae. In the last-named region, the cells contained abundant starch and evidently constituted a storage tissue.

Lang described the mycelium as non-septate with multi-nucleate vesicles similar to those recorded by Janse. Arbuscules and sporangioles were not recorded, but the figures show cells containing hyphae contracted about the nuclei. The gametophyte of *L. clavatum* was regarded as a "total saprophyte with a fungus found living as a symbiont in the tissues probably acting in some way as intermediary."

The variety of structure found in these gametophytes had led Bruchmann to suggest a classification of the genus based on the characters of the prothallia, but Lang's view, that the observed differences of structure were related to habitat and nutrition rather than to systematic affinities is now commonly accepted. These differences, for example, the green assimilating lobes of the gametophyte of *L. cernuum*, and the capacity to produce chlorophyll when exposed to light exhibited by those of *L. selago*, doubtless affect the mode of nutrition, but they have little influence upon fungus invasion, which occurs indifferently in all species.

During recent years, the gametophytes of a number of other species of *Lycopodium* have been discovered, and all agree in general structure with one or other of the types exemplified by *L. cernuum*, *L. phlegmaria*, and *L. clavatum*. Holloway (1920) and Chamberlain (1917) have described those of various New Zealand species. Discussing the profuse fungus infection in the median zone of the subterranean prothallus of *L. volubile*, Chamberlain observed that

> The foot of the sporophyte is strongly haustorial, and the cells surrounding it have some starch but very little protoplasm or other visible contents; consequently the food supply must come largely from the fungus region and must be in a liquid condition even at a considerable distance from the foot.

Spessard (1917) and Stokay and Starr (1924) have given accounts of thalli and sporelings of several American species, and Degener (1924) has recorded the discovery of several hundred gametophytes of *L. obscurum* in a single station. The latter author has also recorded the presence of thousands of prothalli and young sporophytes of *L. cernuum* in volcanic crevices in Hawaii under conditions of abnormal heat. None of these recent observers has devoted special attention to the details of fungus infection.

Reviewing the somewhat scanty data derived from artificial culture of prothalli, it seems legitimate to conclude that, in general, the spores of those forms with green aerial prothalli germinate readily and are entirely or comparatively independent of fungus infection. Those of species with colourless saprophytic prothalli have either not germinated at all under artificial conditions, or have done so with extreme slowness. All the records point to a dependence of germination upon fungal invasion very similar to that in Orchids.

As might be expected from their size and subterranean habit, the prothalli of *Lycopodium* are not readily discovered in nature, and information as to the exact conditions required for their development is still relatively scanty. Ordinarily, they have been found growing sporadically in localised areas identified by the presence of young sporophytes. It may be hoped that recent discoveries on a relatively large scale may lead to fresh information bearing on the exact relation between development and fungus infection, and provide some indication of the specificity or otherwise of the endophytic fungi.

The sporophyte of *Lycopodium* has commonly been described as free from fungus infection and there is at present no record of the formation of ordinary mycorrhiza by any species. In view of the existence of what is probably an obligate relation in the gametophytic phase, the subject is of some interest, and there are observations of possible significance in the literature. Bruchmann (1874) investigated the anatomy of several species of *Lycopodium* without making any record of fungus infection, but in *L. inundatum* he described a tuberous development of the young stem, and the formation of a peculiar tissue—"Polstergewebe"—characterised by the separation of the cells, and the deposition of granular material between them. This tissue appears in the basal part of the young sporophyte and sporadically in the stems of the older plants, and Bruchmann believed that it functioned in relation to the storage of water. Kühn (1889) made similar observations on *L. inundatum*, noted the presence of fungus mycelium imbedded in the slimy material between the cells, and observed that the hyphae could become intracellular—"Das Polstergewebe von *Lycopodium inundatum* enthalt ebenfalls eine Pilzinfection." Goebel (1887) likewise reported fungus infection in the sporophyte of this species of *Lycopodium*. Kühn believed that the production of intercellular slime was directly related to the presence of endophytic mycelium, but was uncertain as to its exact origin, whether from the cell walls or those of the hyphae. Treub (1889–1890), investigating the embryo and young plant of *L. cernuum*, observed the presence of mycelium similar to that in the prothallus—"La plantule du *Lycopodium cernuum* est habitée de même par un champignon, sans doute le même qui se trouve dans le prothalle." The tissue of the embryonic tubercle or protocorm in this species is differentiated into a central mass with large intercellular spaces and a parietal investing layer. At an early stage, the fungus penetrates these intercellular spaces; in older protocorms, the mycelium appears to fill all the free space between the cells and serves to accentuate the contrast between the two kinds of tissue. The central cells retain their nuclei but do not possess starch which is present in the cells of the parietal layer. The young plants of *L. cernuum* produce root tubercles in which the constituent cells become rounded off, leaving intercellular spaces which also, at an early stage, undergo invasion by the mycelium of the endophyte.

The existence of a causal relation between infection and tubercle formation in this species of *Lycopodium* clearly requires investigation. Treub did not put forward any final views respecting the physiology of infection—"Il est hors de doute que l'endophyte ne fait pas de mal aux plantules. Au contraire on reçoit l'impression qu'il y a un certain mutualism entre le tubercle et le champignon."

To elucidate the exact nature of this relation, he pointed out, would require long and patient researches. In spite of Frank's views he evidently regarded the whole subject of the association of vascular plants with fungi other than parasites as still within the region of hypothesis. The rôle assigned by Frank to the root fungi of trees he accepted as possible for the saprophytic prothalli of *Lycopodium* spp., but improbable for the prothallus and young sporophyte of *L. cernuum*, since the latter prefers soils markedly deficient in humus.

With regard to other genera of the Lycopodiales, Bruchmann (1897) reported fungus infection in the roots of *Selaginella spinulosa*, while Stahl (1900) recorded it as absent from those of *S. helvetica*, correlating this observation with the abundant development of root hairs in the latter species. In the rhizome of *Psilotum triquetrum* mycorrhiza was recorded by Solms Laubach (1884), who also observed and figured invasion of the vegetative bulbils by a mycelium—"dont les branches commencent à former les pelotons des hyphes qui se trouvent toujours dans le tissu de la plante." It was noted also by Janse (1896–7), and by Bernatsky (1899), who attempted to isolate the endophyte and extracted a form which he regarded as a *Hypomyces*.

Later, Shibata (1902) undertook a cytological study of the mycorrhiza of *Psilotum* and described it in detail. He noted that the intracellular mycelium underwent complete disintegration by the digestive activity of the host cells and recognised a differentiation of the infected tissue to "Pilzwirthzellen" and "Verdauungzellen" similar to those of Orchids. The nuclei of infected cells showed signs of great activity, and were believed to directly influence both the digestion of mycelium and the production of an amyloid-like material in which the hyphal residues became imbedded. The intracellular mycelium was of the non-septate type and produced vesicles containing fatty substances.

A gametophyte believed to be that of *Psilotum* was described by Lang (1901). The prothallus was found embedded in humus in the neighbourhood of the sporophyte on the stem of a tree fern. It was a small tuberous body about a quarter of an inch in diameter, without chlorophyll, but developing a faint green colour in the exposed tissues. The external layer and the central tissues were free from mycelium; between them, a peripheral zone of cells showed profuse intracellular infection by fine mycelium with vesicles of the usual type, bounded internally by a layer of cells marked by a somewhat unusual development of intercellular vesicles.

The prothallus of *Tmesipteris*, described by Holloway (1917 *b*), and by Lawson (1917, 1918), is also a non-chlorophyllous saprophytic structure showing endotrophic fungal infection of the usual kind.

Ophioglossales.

The three genera belonging to this group are included in the single family Ophioglossaceae. In general, the sporophytic tissues are of simple type. The rather fleshy roots are unbranched or very sparingly branched, and are otherwise characterised by the complete absence of root hairs and the development of typical mycorrhiza in the middle region of the cortex. Russow (1872) recorded fungus infection in mature roots of *Ophioglossum vulgatum*. Bruchmann (1904) described it in the same species, and Campbell (1907) noted it in *O. pendulum* and *O. moluccanum*. Subsequently it was found in other species, notably in the remarkable reduced form *O. simplex* (Bower, 1904).

In transverse section, the roots of *Ophioglossum* show a broad parenchymatous cortex with a zone of typically infected cells towards the middle region.

In *Botrychium*, Grevillius (1895) recorded root infection in twelve species, Kühn (1889) described it in the Moonwort, *Botrychium lunaria*, and Bruchmann (1906) observed it in roots of young plants of the same species, noting that its presence at this stage associated with slow leaf development implied a saprophytic habit for the early stages of growth. Holle (1875) observed that the stronger roots of *Botrychium* were free from infection, which was confined to those showing diarch structure. Marcuse (1902), commenting on the mycorrhiza of *Botrychium lunaria*, remarked that it differed from those of most mycorrhizal plants in that the starch content of the infected cells showed no diminution after entry of the mycelium.

In *Helminthostachys*, Farmer (1899) recorded mycelium in the first three or four roots of the sporophyte, and its absence from those which developed later.

It may be concluded, therefore, that endotrophic mycorrhiza occurs generally in all members of the Ophioglossaceae, but is somewhat inconstant in appearance. Up to the present, this inconstancy has been related to the age of the plant and to that of individual roots rather than to factors in the external environment.

The prothalli in all known members of the family are of the subterranean type familiar in *Lycopodium*. In certain species, traces of chlorophyll have been observed; otherwise they are small cylindrical or tuberous structures, subterranean in habit, quite destitute of chlorophyll, and showing extensive infection by an endophytic fungus. As in *Lycopodium*, knowledge of the gametophytes has been derived partly from prothalli found in nature, partly from artificial cultures. The first to be described, that of *Botrychium lunaria*, was found by Hofmeister (1862). Subsequently, those of various species of *Ophioglossum* were discovered by Mettenius (1856), Campbell (1895, 1907), Lang (1902) and Bruchmann (1904); those of *Botrychium* by Campbell (1895), Jeffrey (1898), Lyon (1905) and Bruchmann (1906); and that of *Helminthostachys zeylanica* by Lang (1902).

Campbell described the gametophytes of *Ophioglossum moluccanum* as from 5 to 10 mm. long, Mettenius those of *O. pedunculosum* as varying from 1·5 lines to 2 inches in diameter. In all species they are usually thickened to form a tuberous swelling at the basal end, and exhibit extensive intracellular infection by mycelium in all but the apical regions. In that of *O. pendulum*, described by Lang (1902), the mycelium was noted as closely resembling that observed by Janse (1896–7) in roots of sporophytes of the same species. In the young prothallus, mycelium occupies the superficial cells at the base and extends into the central tissues of the lower half, avoiding the meristematic tissue and that bearing reproductive organs. The prothalli of *O. pendulum* were found growing in humus among the leaf bases of epiphytic ferns, and probably persist for several years. Infection is conspicuous and the mycelium suffers intracellular digestion. Spores of this species and of others germinate readily in artificial cultures, but have not been observed to develop beyond the 3-celled or 4-celled stages lacking infection. Following his own observations and experiments, Campbell observed—"It is pretty certain that the association with the fungus is a necessary condition for the further development of the endophyte."

The prothallus of *Botrychium* is infected in a similar manner.

Both in *B. virginianum* and in *B. lunaria*, infection extends throughout the greater part of the central tissue leaving a peripheral region, including the meristem on the upper side, free from infection. As in *Ophioglossum*, it probably takes place at an early stage of development (Campbell, 1911). It is reported that invasion by the endophyte is followed by disappearance of starch and accumulation of oil, the latter not readily soluble in alcohol.

The gametophyte of *Helminthostachys zeylanica* has been described by Lang (1902). As in the other genera, mycelium is invariably present showing a distribution similar to that in *Botrychium*. Attempts to germinate spores were unsuccessful, and it may be surmised that the presence of the endophyte is one of the factors essential to success.

The character of the mycelium found in the gametophyte is similar throughout the group. The intracellular hyphae are variable in size and in form, in some cells swelling out to sac-like vesicles, in others giving rise to spherical structures recalling the sexual organs of members of Peronosporaceae. From this similarity, Jeffrey (1898) concluded that the endophyte of *Botrychium* showed affinities with the genera *Pythium* and *Completoria*.

There is no mention of "arbuscules" or "sporangioles" in the literature, although the intracellular mycelium apparently undergoes digestion, and in other respects the endophyte resembles the ubiquitous "Phycomycete type" recorded by Peyronel. In respect both to the obligate character of the association in the gametophyte phase and the identity of mycelium in gametophyte and sporophyte generations more information derived from experimental cultures is required. From the known facts, it is reasonable to conclude that an obligate relation with specific endophytes has been evolved in the gametophytes of all genera, and that the young roots of developing sporophytes are subject to infection by the same fungus. Whether such infection is a necessary condition for the normal development of roots is not known.

Filicales.

In the Filicales, mycorrhiza formation in the sporophyte, and a corresponding type of fungus infection in the gametophyte, is found regularly in the family Marattiaceae and sporadically in other groups. It is of rare occurrence in the Leptosporangiate Ferns and has not hitherto been put on record for the Polypodiaceae (see p. 185).

Marattiaceae. For the six genera of ferns included in the Marattiaceae, there are a number of isolated records respecting root infection in the sporophyte, but practically no experimental data. Russow

(1872) summarised his comparative observations on members of this group and the Ophioglossaceae as follows:—

In den unterirdischen Wurzeln fallen die zwei bis drei inneren Lagen der Aussenrinde vor den übrigen in den Augen durch den Inhalt ihrer Zellen, der aus Zussammengeballten, schwach gelblich tingirten zum Theil durchscheinenden, zum Theil grumösen Massen besteht, die sich auf Zusatz von Jod dunkel schmutzig-gelb farben in den Wurzeln der Ophioglossen kommen in den entsprechenden Zellen ähnliche Conglomerate vor, die durch eingedrungene Pilzfäden vernsacht zu sein scheinen; bei den Marattiacen waren keine Pilzfäden wahrzunehmen.

Kühn (1889) described endophytic mycelium in the roots of *Angiopteris evecta*, *Kaulfussia aesculifolia*, and *Marattia alata*, but sought in vain for evidence of infection in *M. fraxinea*. He claimed also to have isolated from roots of *M. alata* an endophyte which spored in pure culture. No proof of identity was obtained and, from the methods employed, it is not unlikely that the fungus in question was a member of the epiphytic flora of the roots. Stahl (1900) failed to observe infection either in *M. alata* or in *M. fraxinea*, but recorded it as constant in *Angiopteris evecta*. Gallaud (1905) figured the mycorrhiza of *Angiopteris durvilleana*, describing it as of similar type to that of *Arum maculatum* (cf. Fig. 12 *a*, p. 59).

West (1917) published an account of his observations on a number of marattiaceous species, and his main conclusions are included in the following brief summary. In *Angiopteris henryi*, *Kaulfussia aesculifolia*, and *Marattia cooperi*, the endophyte was found regularly in the primary roots and earlier adventitious roots but was inconstant in appearance in those formed later. It was usually absent from the aerial parts of the roots, from a proportion of the mature roots of all species, and from all the roots of some of the plants examined. This inconstancy may explain the failure of Kühn and Stahl to find infection in certain species of *Marattia* and in *Danaea alata*, and also the statements made by Campbell (1911) and Charles (1911) respecting the distribution of the endophyte in older roots. There is no evidence that the incidence of infection is correlated in any way with soil conditions.

The distribution of mycelium is inter- and intracellular; vesicles and arbuscules are formed (cf. Gallaud, Figs. 13, 34, 35), the latter showing every stage of degeneration due to the digestive activity of the root cells. An unusual feature recorded by West is the presence of thick-walled resting spores believed to be identical with those observed by Kühn in roots of *Kaulfussia*. Attempts to isolate the fungus met with no success. On the morphological characters, West believed its affinities were with members of Peronosporaceae, and he placed it in a new genus *Stigeosporium*, naming it, by reason of the habitat, *S. marattiacearum*.

The roots of *Danaea alata* and *D. nodosa* also form mycorrhiza, the endophyte being apparently not *Stigeosporium* but a distinct form. In view of the absence of reproductive bodies no attempt was made to determine the affinities and systematic position of these fungi.

The gametophytes of members of Marattiaceae are green thalli resembling those of the Leptosporangiate Ferns but more massive in structure and longer lived. Campbell (1911) recorded an endophyte similar to that found in the prothalli of members of Ophioglossales as present in almost every case in those of *Angiopteris evecta* and *Kaulfussia aesculifolia*. A similar fungus was known to occur in *Marattia douglasi* and was presumably also present in prothalli of other species of this genus.

In the central tissues, intracellular infection was observed to be associated with the disappearance of starch and degeneration of plastids in uninfected cells. The irregular vesicular swellings which suffer digestion in the prothalli of Ophioglossaceae are said to be absent. Experimental data are scanty but there appears to be no evidence that spore development is in any way bound up with infection.

In both groups of Eusporangiate Ferns, the biologic relations appear to be of the usual kind. In its mode of entry the endophyte behaves as a parasite, but the invaded cells show no symptoms of injury and the attack is confined to the absorption of starch and other non-living cell constituents, a proportion of which become once more available to the host by subsequent digestion of intracellular mycelium. There is, at present, no experimental data bearing on the possibility of nitrogen fixation by the endophytes, nor any which throws light upon their reaction to the organic constituents of the humus soils in which many of these prothalli grow.

Stahl (1900) drew a comparison between the Ophioglossaceae with their reduced root systems, absence of root hairs and regular endophytic infection and the Marattiaceae with—as he believed—a more efficient water economy and less frequent infection. In respect to the latter group he held that they occupied a position intermediate between the regularly infected Ophioglossaceae and the ordinary Ferns (Polypodiaceae) which were reported free from mycorrhiza.

Filices.

In this group—the Ferns in the popular sense—the formation of mycorrhiza is an uncommon phenomenon, a curious and at present unexplained fact in view of the distribution of the members on woodland and other humus-rich soils.

Janse (1896–7) observed root infection in *Cyathea* sp. in Java but found none in the epiphytic *Asplenium nidus-avis.* Stahl (1900) noted the absence of mycorrhiza from *Osmunda regalis*, the Hydropterids, and many members of Polypodiaceae; namely, *Aspidium filix-mas, A. lobatum, Asplenium filix-femina, A. viride, Pteridium aquilinum* and *Polypodium vulgare.* Frank (1887 *b*) had already observed that *Aspidium thelypteris* was uninfected. In view of the absence of any positive record of mycorrhiza in members of Polypodiaceae and the existence of more than one negative observation for *Aspidium filix-mas*, it is of interest to place on record its appearance in roots of this fern from the neighbourhood of London. In this, at present, unique record of mycorrhiza formation in the family Polypodiaceae, the mycelium is concentrated in a single layer of cells in the inner cortex and the intracellular mycelium shows the usual characteristic features. The relation is evidently entirely different from that in ordinary parasitic invasion by mycelium and shows the characters and features invariably associated with endotrophic mycorrhiza (Pl. VII, Figs. 59, 60).

It is probable that endophytic mycorrhiza occurs in the roots of members of Polypodiaceae more commonly than has been supposed. I am indebted to Dr E. McLennan for the following unpublished note recording its occurrence in Bracken (*Pteridium aquilinum*):—

Roots of the Bracken fern (*Pteridium aquilinum*), obtained from Castlemaine, Victoria, Australia, when sectioned and examined microscopically, showed a typical endophytic mycorrhiza. It was noticed that entrance to the root was effected through the root hair, the travelling hyphae carrying the 'infection' horizontally through the outer and middle cortex, while the digestive zone occurred only in the endo-cortex. Arbuscules and sporangioles were observed in the two or three cortical cell-layers immediately adjacent to the endodermis. The soil from which the roots were collected was a heavy clay.

In the gametophyte generation of the Leptosporangiate Ferns, there are isolated records of infection, but no evidence of the existence of a specialised relation with the endophyte. In *Osmunda cinnamomea*, Campbell (1908) noted that many prothalli contained an endophytic fungus resembling that in *Ophioglossum* and *Botrychium*. He also observed a similar type of infection in prothalli of four species of *Gleichenia* from widely separated geographical regions, namely, *G. polypodioides*, *G. dichotoma*, *G. laevigata* and *G. pectinata*. He concluded—"that an endophytic fungus is normally present in the green prothallia of several Marattiaceae, Osmundaceae, and Gleicheniaceae, and it is highly probable that further research will show similar fungal endophytes occurring in the prothallia of many other ferns." Campbell also put forward the view that the presence of the endophyte may have been an important factor in the evolution of the saprophytic, subterranean gametophytes of the Ophioglossaceae from green holophytic forms resembling those of Marattiaceae.

Commenting on the evidence of infection in Fern prothalli in general, Campbell has noted the possible occurrence of a series leading from complete saprophytism and dependence upon infection, as in Ophioglossaceae, to a more or less casual type of infection, as in the green prothalli of Marattiaceae and the families of Leptosporangiate Ferns.

With the exception of a few experimental observations on the germination of spores with and without the endophyte, there is at present no experimental evidence upon which to base an opinion as to the precise significance of infection in Ferns. There is evidently a close parallelism with the condition in Lycopods, both in respect to the inconstancy of infection in the sporophytes and its invariable association with a non-green saprophytic gametophyte stage.

The evidence put forward by Stahl (1900) in support of his theory of nutrition in mycotrophic plants has been subjected to criticism in more than one instance in the present review. His attempt to correlate fungus infection with economy of water exchange in Ferns was criticised by Bower (1908), who pointed out that there is less evidence of this in *Cyathea*, which forms mycorrhiza, than in *Asplenium nidus-avis* and *Osmunda regalis* which do not. Bower finds little evidence among Ferns that the mycorrhizal habit is an effective source of organic nutrient supply. If it were, he adds, it might be expected that examples would occur showing vegetative reduction and loss of chlorophyll, whereas, with rare exceptions, e.g. *Ophioglossum simplex*, this is not the case.

In Bower's opinion, the facts do not bear out the general assumption that mycorrhizal symbiosis, as seen in certain Pteridophytes, is directly associated with vegetative reduction of the infected sporophyte as a whole. In this respect the section of *Ophioglossum* known

PLATE VII

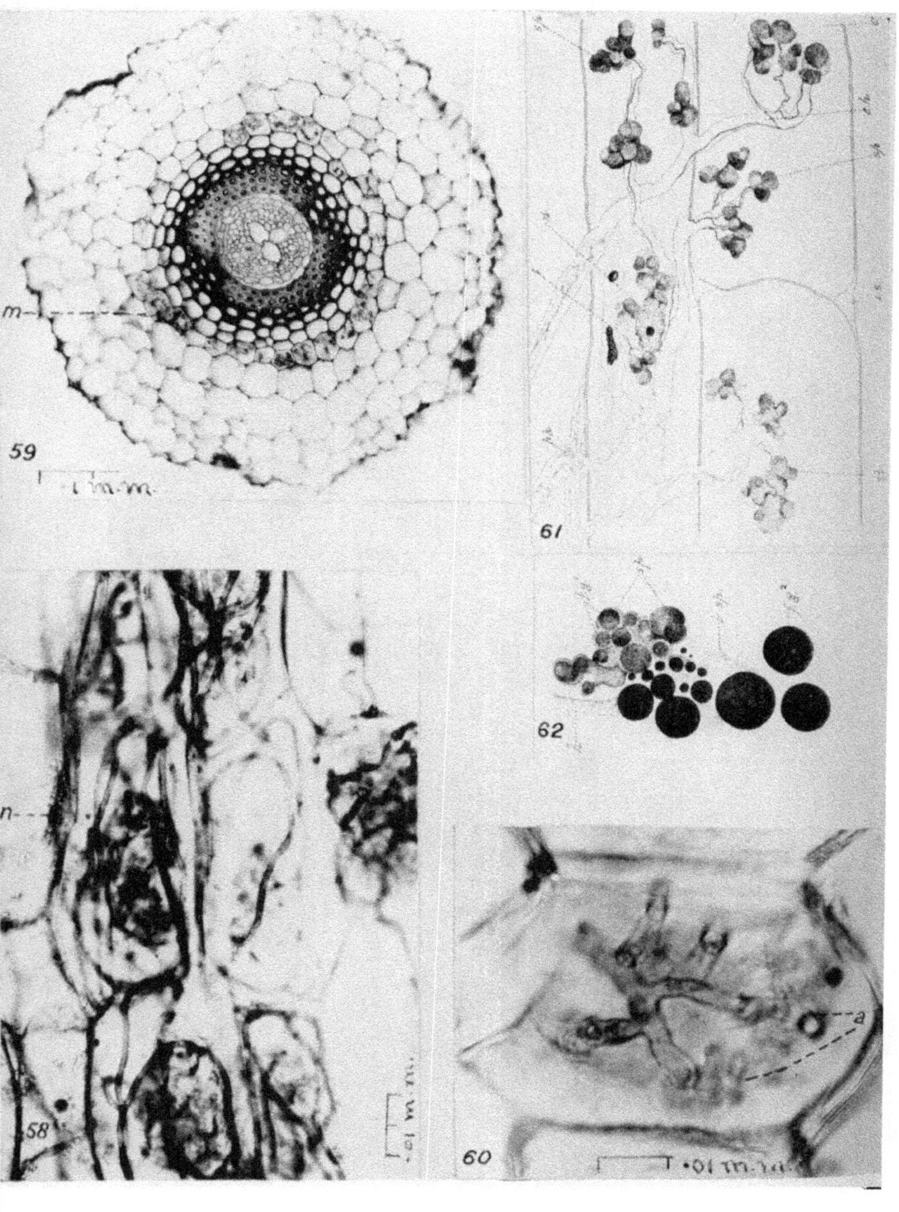

as *Ophioderma* is exceptional, and is regarded as providing a series in which mycorrhiza has become effective as a partial substitute for chlorophyll nutrition, while reduction of the vegetative system has actually followed as a consequence. *Ophioglossum simplex* represents an extreme case showing complete disappearance of the sterile lamina of the leaf.

The family Psilotaceae has also been cited as one in which a tendency towards reduction of the vegetative organs points to the efficiency of mycorrhiza as a nutritive mechanism.

EXPLANATION OF PLATE VII

Fig. 58. *Aneura* sp. Cells of thallus showing infection by a mycorrhizal fungus of the *Rhizoctonia* (Orchid) type. *n*, cell nucleus.

Fig. 59. *Aspidium filix-mas*. Transverse section of root showing development to mycorrhiza. *m*, infected cells.

Fig. 60. Single cell from mycorrhizal layer shown in Fig. 59, more highly magnified. Main branches of intracellular mycelium with arbuscules, subsequent to digestion. *a*, arbuscules.

Fig. 61. *Lolium temulentum*. Cells from the innermost zone of the root cortex, prior to bursting of the sporangioles. At this stage, the cell contents include the nucleus, cell cytoplasm, shrivelled hyphae, and intact sporangioles enclosing the fat about to be liberated into the cell lumen. *n*, nucleus; *pp*, protoplasm; *i.s.*, intercellular space; *e.h.*, empty hyphae; *sp*, sporangioles; *c*, cell of the innermost zone of the cortex. Original × 1050. (From McLennan, *Annals of Bot.* 1926.)

Fig. 62. Portion of a cell similar to those shown in Fig. 61 at the exchange stage, showing intact and burst sporangioles. *h*, hypha; *sp.*, intact sporangiole; *b.sp.*, burst sporangiole; *f.g.*[2], free fat globules; *f.g.*[1], fat globules seen through wall of intact sporangiole. Original × 1260.

(For text reference to Figs. 61 and 62, see Chapter XI.)

CHAPTER X

TUBERISATION

Tuberisation: The association of root nodules or tubercles with fungus infection—Historical—Nodules of leguminous plants: early views—Root nodules of *Alnus, Eleagnus, Myrica, Ceanothus*: early observers; Bottomley; Spratt—The root tubercles of Cycads—Root nodules of *Podocarpus* and other Conifers: Spratt; McLuckie; Yeates. The tuberisation theory: Bernard; Magrou.

OBSERVATIONS associating the presence of fungus mycelium with the formation of nodules or tubercles are not uncommon in the literature of root infection and merit brief notice, albeit many of the earlier records betray inaccuracies due to faulty observation and still more faulty technique.

Malpighi (1679) figured small nodules on the roots of legumes, mistaking them for insect galls, and similar observations were made by Duhamel (1758). Meyen (1829) noted the root tubercles of Alder and believed them to be parasites similar in habit to members of the Orobanchaceae, although of more lowly development. Schacht (1853, 1860) also observed these conspicuous outgrowths on Alder roots, regarding them first as normal and afterwards as abnormal outgrowths, but he made no attempt to explain their presence. Woronin (1866) believed that the organism present belonged to the genus *Schinzia*, founded by Nägeli for a mycelium observed in Iris roots; he named it, accordingly, *S. alni*. Later, following a paper by Gravis (1879) on the same subject, he reinvestigated the structure of young nodules, and found in the cells an organism believed to be identical with his previously described *Plasmodiophora brassicae* in Turnip. Woronin concluded that both a Myxomycete and a filamentous fungus were constantly present in the nodules of Alder, a belief which Gravis (1885) endorsed in a subsequent paper.

Warming (1876) noted similar "galls" in *Hippophaë, Eleagnus* and *Shepherdia*, and discovered in all cases an organism resembling the "*Plasmodiophora*" of Woronin. Moeller (1885), after repeating once again the observations on Alder, named the endophyte *P. alni*, but Woronin adhered to his original view that a filamentous fungus was also present.

The confusion caused by all these conflicting views was partially cleared up by Brunchorst (1886), who, by the use of improved methods, showed that the so-called "plasmodia" were actually the cell protoplasts, imbedded in which were fine filaments believed to be fungus hyphae. The bodies mistaken for spores by Woronin and Möller he held to be sporangia.

Brunchorst referred the fungus in Alder nodules to a genus distinct from *Schinzia*, proposing for it the new generic name, *Frankia*. Unfortunately he rejected the well-known specific designation and renamed the organism *Frankia subtilis*; he regarded the mycelia present in the nodules of the various other plants mentioned as belonging to the same fungus. Subsequently, Woronin accepted Brunchorst's views on Alder nodules in the main, though still differing in respect to the interpretation of certain details of structure, and incidentally, caused further confusion by giving the name *Frankia brunchorsti* to the organism present in *Myrica gale*.

Equal disagreement existed in respect to the significance of these structures. In Alder, Ward (1887) believed that the mycelium was responsible for certain changes appearing in the cytoplasm of the host cells. Frank (1887*c*) rejected the view that the nodules were due to parasitic invasion, and expressed the opinion that they were normal organs functioning for the transitory storage of proteins. In view of the existing confusion he further suggested that the names *Schinzia alni*, *Plasmodiophora alni*, *Frankia subtilis* and *Schinzia leguminosarum* should be deleted from mycological literature!

Weber (1884) examined the root nodules formed by the Toad Rush (*Juncus bufonius*), and referred to certain earlier observations by Magnus, who had described a mycelium present in the roots of this species and also in those of *Cyperus flavescens* under the name of *Schinzia cypericola*.

The structures produced by the Toad Rush resemble miniature Potato tubers (Fig. 63). Formed only by certain individuals, it is at present an open question if they should be regarded as a varietal character or as a response to soil or other external conditions. Their precise morphological nature is not clear from the published accounts, while the possible existence of a causal relation with fungus infection requires experimental investigation.

Frank (1891) expressed the opinion that the root nodules of Alder and of members of the Leguminosae were biologically related to mycorrhiza. Since they appeared to be structures morphologically distinct from roots, he proposed that they should be named "Mykodomatien." The exact nature of the endophyte was in some doubt. While expressing the view that the filaments resembled those of a bacterium belonging to the *Leptothrix* group, he evidently inclined to the view that they were fine hyphae. In the case of certain legumes, e.g. *Phaseolus vulgaris*, he held that the nodule organism was purely parasitic. In the same paper Frank (1891) elaborated his conclusions in respect to "Pilzsymbiosis" in general, enumerated the families in which mycorrhiza had been observed, and noted that the true nature of the "Rhizobia" was, in most cases, still unknown. Much controversy took place at this time and later respecting the real nature of the "Infectionsfäden" in the tissues of leguminous nodules.

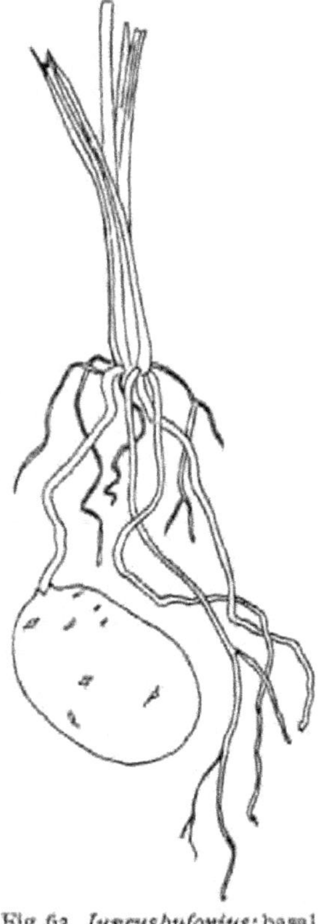

Fig. 63. *Juncus bufonius*; basal part of a plant showing root system and a single tuber. ×7.

Schneider (1892) made observations upon American "Rhizobia" and published a scheme of classification to include the forms observed in different species. He contributed his own views to the controversy respecting the exact nature of the nodule organisms, which, by that time, had been described by Woronin as "Bakterien," by Brunchorst as "Bakteroiden," by Ward (1889) as "spores" or "gemmules" and by Beijerinck (1890) as a specific bacterium, *Bacillus radicicola*. Subsequently he published an historical survey of the history of leguminous nodules and the literature concerning fixation of atmospheric nitrogen by plants, while Bottomley (1907), contributed to that dealing with the morphology of root nodules in general.

When Hellriegel and Wilfarth brought to a close a long series of experiments dating back to 1862 and put forward their new hypothesis respecting the nitrogen-assimilating powers of leguminous species, they established experimentally two essential facts. One, the direct relations between nodule formation and the ability of legumes to utilise atmospheric nitrogen: the other, the dependence of the nitrogen-fixing capacity on the activities of organisms present in the nodules.

It is outside the scope of the present review to attempt a survey of the voluminous literature dealing with the root nodules of leguminous plants. Some mention of the subject is called for in an historical account of work on root infection, because the earlier observers commonly attributed the formation of root nodules—whether by members of Leguminosae or of other groups—to endotrophic fungus infection similar to that in mycorrhizas. Of recent publications dealing with various aspects of the subject may be mentioned papers by Spratt (1919), Lohnis and Hansen (1921), Bewley and Hutchinson (1920), and Wright (1925).

Nodules very similar in appearance to those of legumes are formed not only by Alder, but also by species of *Eleagnus*, *Ceanothus*, and *Myrica*. Hiltner (1896) had reported the presence of bacteria in the root nodules of Alder and *Eleagnus*, and demonstrated experimentally that young Alders without nodules did not thrive in soils poor in nitrogen, whereas inoculation by the appropriate organism was followed by nodule formation accompanied by greatly increased growth. He had noted also that the production of nodules was more active in poor soils, and inferred that the numbers produced by Alder under conditions of nitrogen deficiency could be used as a measure of the activity of the tree in respect to nitrogen fixation. The nodules in the roots of *Ceanothus americanus* were examined by Beal (1890) and Atkinson (1892), the latter of whom named the organism present *Frankia ceanothi*. Later, Bottomley (1915) re-investigated them and, in the light of his own work on similar structures in other plants, reported that they represented modified lateral roots infected by a bacterium belonging to the *Pseudomonas* (*Bacillus*) *radicicola* group.

The work of Shibata (1902) on *Alnus incana* and *Myrica gale* has already been noted. Although he inferred infection by fungus mycelium in the roots of both species, this author noted that in *Alnus* the structure of the thread-like "hyphae" was bacterial in character, while in *Myrica* he suggested that the endophyte was related to *Actinomyces* rather than to the Fungi proper.

From root nodules of the same species, Peklo (1910) isolated an organism which he identified as a species of *Actinomyces*, but attempts to reproduce nodule formation by means of inoculation were not attended by success. More recently, Dufrénoy (1920) has recorded *Actinomyces* sp. as an endophyte in roots and shoots of *Adenostyles albifrons* in the Pyrenees, and expressed the view that shoot infection of a similar type likewise occurs in members of Compositae, Orchidaceae and Ericaceae, although further work is required to establish this claim.

The nodules of *Myrica* were subsequently investigated by Bottomley (1912) who reported infection by a nitrogen-fixing organism belonging to the *Pseudomonas radicicola* group. Old nodules and the basal regions of younger ones were subject likewise to occasional infection by mycelial fungi.

The curious coralloid roots produced by all genera of living Cycads have often attracted attention. They were studied by Reinke (1873), Schneider (1894), Life (1901) and Zach (1910). According to Spratt (1915), the initial stimulus to their formation is due to infection by *Pseudomonas radicicola*. Later, they suffer multiple invasion by other micro-organisms—invariably by species of *Azotobacter*, frequently also by species of *Anabaena* that give rise to the characteristic algal zone.

Mention has been made of the nodular roots formed by *Podocarpus*. Experimental enquiry convinced Nobbe and Hiltner (1899) that these structures were actively concerned in the fixation of atmospheric nitrogen. Thus, they found it impossible to cultivate young plants of *Podocarpus* in quartz sand lacking combined nitrogen in the absence of the organism responsible for nodule formation, whereas plants possessing nodules made healthy growth during five years' cultivation in like conditions.

Shibata (1902), working on two Japanese species of *Podocarpus*, identified what he believed to be the mycelium of a mycorrhizal fungus in the root tissues; he described in detail the cytological changes brought about by the digestion of the intracellular mycelium, and demonstrated the presence of active proteolytic enzymes by means of glycerine extracts of the nodule tissues.

The nodules produced by *Podocarpus* and other members of the Podocarpaceae are characteristic in appearance and arrangement, giving a curious "beaded" appearance to the affected roots. Their morphology and physiological significance was studied by Spratt (1912) who reported the presence of nodules of similar type in *Podocarpus*, *Microcachrys*, *Dacrydium*, *Saxegothaea*, and also in *Phyllocladus*, a genus whose systematic affinities have aroused some controversy. In all the genera examined, they were reported to be perennial structures formed by the modification of lateral roots following infection by nitrogen-fixing organisms identical in structure and behaviour with the strains of *Pseudomonas radicicola* extracted from the root nodules of leguminous plants and also from those of certain non-leguminous species. The majority of the cortical cells undergo some structural modification and give rise to a characteristic water-storage tissue, in certain cells of which the endophyte may remain dormant during the winter. In a subsequent paper, Spratt (1919) reviewed the whole subject of nodule formation, and summarised the experimental evidence on which were based the conclusions reached by herself and by Bottomley, in respect to the identity of the nodule organism in all the cases just cited.

If the views of these observers are accepted, it may be assumed that the root nodules of *Alnus*, *Myrica* and members of the Eleagnaceae, Ceanothaceae, Cycadaceae, and Podocarpaceae play a rôle in the economy of their vascular hosts similar to that securely established for the corresponding structures provoked by the presence of the nodule bacteria in roots of legumes.

In a series of papers entitled "Studies in Symbiosis" McLuckie (1923 *a*) has recently contributed his observations on the nodular roots of certain Australian species of *Podocarpus* and *Casuarina*. He has confirmed the opinion expressed by Spratt that those of *Podocarpus spinulosa* and *P. elata* are due to infection by a bacterium showing many points of resemblance with *Pseudomonas radicicola* although not certainly identified with that species. In pure culture outside the plant, the organism present is reported to fix atmospheric nitrogen at the rate of 6 to 7 mg. of nitrogen per 100 c.c. of nutrient solution in 21 days. In *Podocarpus*, the cortical cells of the main roots and those of some of the nodules were observed likewise to contain mycelium of a mycorrhizal type.

The root nodules of *Casuarina cunninghamia* were reported by McLuckie to be due to infection by a similar type of nitrogen-fixing bacterium to that isolated from *Podocarpus*. In another species, *C. equisetifolia*, from coral islands near Java, the roots bear structures resembling the nodules of Alder. These were examined by Miehe (1918) and found to consist of closely compacted masses of freely branched, shortened rootlets, in certain cortical cells of which fungal hyphae were present. The tubercles were reported on the roots of all individuals growing in a natural habitat but were not observed on those in cultivation in Europe. They were named "*Rhizothamnion*" by Miehe who believed that they represented a typical case of beneficial symbiosis in which the fungus partner functioned by converting nitrogenous materials in the humus into forms directly available to the vascular host. In view of the possibility of mistaking the zoogloea threads of a bacterial endophyte for fungal hyphae, and the uniformity of all recent records associating a nitrogen-fixing organism of the *Pseudomonas radicicola* type with root nodules, confirmation of Miehe's observation on *Casuarina* would be welcomed.

Yeates (1924) has made a further contribution to the subject of root nodules in conifers, based upon observations made on species indigenous to New Zealand. He reports nodule formation by "all the New Zealand pines" with the exception of two species of *Libocedrus*. He found them also on several members of the Araucarineae including the Kauri Pine (*Agathis australis*), on which they had been noticed previously by Cockayne (1921). In respect to the morphology of the nodules and their function as organs for water storage, Yeates' observations and deductions agree with those of Spratt. In this connection, the former has noted that the size of the juvenile leaves in species of *Dacrydium* and *Podocarpus* is roughly proportional to that of the root nodules, i.e. the forms with larger leaves showing also the greatest development of water-storage tissue. Moreover, the species showing the smallest nodules are likewise those in which the transition from the large juvenile type of leaf to the reduced form characteristic of the adult, is most striking, e.g. *Dacrydium biforme*: "In other words, absence of sufficient water-tissue on their roots has compelled these species to reduce their leaf-surface and so to economize their water supply"—an argument cogent enough in itself although expressed in somewhat teleological form. From his examination of the mycorrhiza of *Cunninghamia*, *Libocedrus*, and *Cupressus*, Yeates is prepared to extend this hypothesis to conifers generally, in the form that root reduction consequent upon mycorrhizal infection is accompanied by parallel diminution in size of the leaves.

His thesis that the Abietineae constitute an isolated case in which absence of endophytic mycorrhiza is correlated with a tendency towards persistence of the juvenile leaf habit, is hardly borne out however by the recent observations of Melin (see Chap. VIII) on the mycorrhiza of Spruce and Pine.

In respect to the characters of the nodule organism in Taxads, the observations of Yeates are markedly at variance with those of Spratt. The vast majority of nodules investigated by the former are reported as showing infection by the characteristic type of mycorrhizal fungus with non-septate mycelium and vesicles; only in less than 1 per cent. of some hundreds of nodules examined did he observe bacteria similar to those figured by Spratt and also by McLuckie for *Podocarpus*. Yeates has concluded that the nodules function primarily as organs for water storage and are also beneficial to the trees by enabling them to draw indirectly upon the organic residues present in the soil through the agency of their mycorrhizal fungi.

Reviewing the cases described, it may be concluded that there is evidence that the formation of nodules by a number of plant species other than legumes is directly related to invasion of the roots by strains of a bacterial organism closely related to *Pseudomonas radicicola*. Whether this conclusion can be extended to include the Podocarpeae and other nodule-forming conifers is at present doubtful, and requires confirmation by means of pure culture experiments on the species concerned. For example, in view of the conflicting nature of the observations made by Spratt and Yeates respectively, it would be desirable to investigate the response shown by *Podocarpus* seedlings to independent infection by the bacteria isolated from the nodules.

On the other hand, evidence is not lacking that tuberisation, in greater or less degree, is often a consequence of fungal infection. The "Knollenmykorrhizen" of Pine and Fir (see Pl. VI, Fig. 50) may be cited as a case in point and in external features these bear some resemblance to the nodular roots of Alder and Cycads. It was, indeed, the constant association of local hypertrophy with endophytic infection by mycelium that led Bernard (1902) to put forward the theory of tuberisation subsequently extended and elaborated by Magrou (1921) and others. A short account will now be given of this interesting and far-reaching hypothesis, with a critical examination of the evidence adduced in support of it.

Laurent (1888) had shown that tuberisation in plants could be provoked experimentally by raising the concentration of the culture medium. Bernard's attention was attracted to the subject during his early observations on the Ophrydeae, in members of which group the formation of tubers is so constant and regular that their morphology can be utilised as a basis of classification. He was struck by the prevalence of the habit among herbaceous perennials in general and, finding no clue in Laurent's work to the stimulus provoking tuberisation under natural conditions, he looked around for a cause "as widespread as the habit." He was familiar with the prevalence and constancy of the mycorrhizal habit in wild perennials and also with Stahl's observation that species producing bulbs and tubers are specially prone to infection. His own researches on members of the Ophrydeae convinced him that a causal relation between fungus invasion and tuberisation existed in members of this group.

So, by more than one line of reasoning, he was led to the view that tuberisation (or its equivalent) and, in general, the perennial habit in herbaceous plants, represents a condition of relatively advanced adaptation to communal life between vascular plants and fungi, and is in itself a direct consequence of such a symbiotic relation. With this theory was bound up an equally speculative hypothesis tracing the evolution of vascular plants from thalloid members of Bryophyta as an indirect consequence of fungal infection and the establishment of the perennial habit in the latter.

The evidence collected by Bernard from observations on *Ophrys* and allied Orchids may be briefly summarised as follows. Seed germination leads to the formation of a small tuberous structure, the primary tubercle, bearing an apical bud and showing profuse infection by mycelium at the basal end. During the first season's growth, the bud produces a few leaf rudiments and a lateral swelling —the first tuber—which is eventually set free in the soil. The primary tubercle dwindles and disappears, while the tuber, still entirely free from fungus infection, gives rise, during the ensuing vegetative season, to a short axis bearing absorptive hairs. The latter quickly become infected from the soil, as may later the tuber itself. The same sequence is repeated annually until the seedling reaches maturity. In the autumn preceding flowering, the roots, produced at the end of September, become infected from the soil and remain in this condition until their death after flower production in the following spring. One or more of the axillary buds then develop to a tuber or tubers, and the plant remains free from infection until the new roots are produced in the early autumn. Thence onwards this regular annual periodicity persists as a normal characteristic of the growth cycle.

Two features were of special interest to Bernard, one, the invariable infection of roots as compared with the complete immunity shown by tubers; the other, the regular annual periodicity exhibited in respect to fungus infection. During August and September there is active vegetative growth without infection; from October, following the invasion of the roots by mycelium from the soil, to June, there ensues a period of tuberisation. In Bernard's view the two series of phenomena are causally related, i.e. fungal infection produces a general "intoxication" of the tissues made manifest, in the embryo, by the formation of the primary tubercle, in the adult plant, by the annual tuberisation of one or more of the axillary buds subsequent to root infection. In short, a brief phase of growth and differentiation with freedom from infection alternates regularly with a longer period of tuberisation following root infection.

Later, Bernard showed that non-infection of the tubers was due to a fungicidal action exerted by the tissues (see p. 78), while he also demonstrated experimentally by his work on *Bletilla* that tuberisation of uninfected embryos could be brought about by raising the concentration of the substrate. Thus, seeds of this Orchid sown in aseptic culture developed slowly to plantlets with slender stems bearing leaves separated by distinct internodes, whereas, seeds germinated with an active strain of the endophyte showed rapid infection of the embryo followed by vigorous development, the resulting seedlings exhibiting marked differences in habit as compared with those lacking infection, e.g. they produced relatively massive axes with short stout internodes and crowded leaves of much larger size than those of uninfected seedlings. The tuberous base of the stem, at first covered by absorbing hairs, soon produced roots; as in the Ophrydeae, the latter suffered infection from the soil, while the swollen stem and protocorm remained free from infection.

Can this theory of tuberisation be applied to Flowering Plants generally? In seeking the answer to this question Bernard extended his observations to plants other than Orchids, e.g. to *Ranunculus ficaria* and the cultivated Potato (*Solanum tuberosum*). In the former, he thought the evidence was confirmatory of a relation similar to that which he believed existed in *Ophrys* and allied Orchids; in Potato, the observed facts were more difficult to explain.

It is well known that healthy potato tubers are free from infection by micro-organisms of any kind. Bernard's observations confirmed this and showed that the roots suffered infection of an irregular and apparently casual kind. Subsequently a fungus, believed to possess affinities with the genus *Mucor*, was isolated from the roots of *Solanum dulcamara*, a common British and European species of the same genus as Potato. At this stage the work was interrupted by Bernard's untimely death, and the identity of the fungus with that endophytic in the roots of the latter was left in doubt.

The investigation has since been resumed by his colleagues at the Pasteur Institute, who have extended the scope of the enquiry and repeated some of Bernard's later experiments. In a recent memoir Magrou (1921) has restated the theory of tuberisation just outlined and marshalled the evidence accumulated by Bernard and his successors. The observations on Potato are of particular interest.

Assuming that the ordinary relation between root infection and tuberisation in the Potato has been replaced by effects due to cultivation, Magrou examined *S. maglia*, a wild Chilean species believed to be a direct descendant of the original ancestor of the cultivated Potato, and also *S. dulcamara*, the Bittersweet. In both species, the observed existence of mycorrhiza was thought to provide confirmatory evidence of a direct relation between infection and tuberisation. The irregularity of infection observed in the cultivated Potato is explained by assuming that the stimulus originally associated with fungus infection has been replaced by others depending upon high cultivation and manuring, "selection" having co-operated to preserve only the individuals which respond in this way. In support of this view, an appeal is made to records made in France in the eighteenth century by Parmentier, who is reported to have observed whole fields of potatoes without tubers, and to the occasional appearance on non-tubering individuals in ordinary crops. It must be noted, however, that an equally good explanation of this phenomenon of "zero-cropping" has been offered on genetical grounds by Salaman (1924) who interprets the absence of tubers as due to the presence of "inhibiting factors" in the genetical constitution of individual Potato plants or strains.

Owing to the scanty and irregular appearance of mycelium in the roots of cultivated Potatoes, *Solanum dulcamara* was used as a source of infection, in Magrou's experiments. Potato seed sown on poor soil from a station occupied by *S. dulcamara* gave seedlings which grew slowly but developed roots with typical endophytic infection. Evidence is offered that two distinct types of plants appeared in these cultures; those in which fungus infection takes place from the soil but the endophytic mycelium undergoes immediate and complete digestion and those in which infection persists giving rise to characteristic endophytic mycorrhiza. Plants of the former type possess "immunity" and consequently do not tuberise; those of the other type develop an abundant crop of tubers (Fig. 64).

Confirmation of the correctness of this interpretation can be provided only by synthetic experiments with "pure cultures" of the organisms concerned. Up to the present, experiments of this kind with Potato have not yielded perfectly satisfactory results. A fungus resembling that described by Bernard was isolated by Magrou from *S. dulcamara* and named *Mucor solani*. This form causes typical infestation of roots of Potato seedlings raised from sterilised seed and is believed by Magrou to be identical with, or closely related to, the original endophyte of the ancestral species.

It has, however, not yet been found possible to provide vigorous proof of a direct relation between root infection and tuber formation by means of strictly aseptic cultures as in the Orchids and *Calluna*, and a final expression of opinion must await the confirmatory evidence supplied by such cultures.

The work on Potato has been supplemented by observations on other plants: two species of *Orobus*, *O. tuberosus*, a perennial species with tuberous rhizomes, and *O. coccineus*, an annual species without a perennating shoot system; *Mercurialis perennis*, the Dog's Mercury, with an elaborate system of rhizomes, and *M. annua*, an annual herb with aerial shoots only. In both species of *Orobus*, infestation of the roots occurs; in *O. tuberosus* it persists as mycorrhiza; in *O. coccineus* the invading mycelium undergoes immediate destruction by the root-cells of the host, i.e. the symbiotic relation is not established and the annual habit persists. In the two species of *Mercurialis* named a similar state of things is reported and figured. In both genera, the difference in habit is believed to be associated with the higher "immunity" possessed by that species in which the mycorrhizal relation is not established (Fig. 64).

Taking into account his own observations, Laurent's demonstration that tuberisation can be provoked by raising the concentration of sugar in the culture medium, and Bernard's experiments proving that in pure culture the Orchid endophytes bring about a rise in the osmotic pressure of the substratum, Magrou considers there is good presumptive evidence for believing that, in some cases at least, the immediate cause of tuberisation is an increased osmotic pressure in the root tissues due to their invasion by endophytic fungi. Magrou's conclusions have been challenged by Peyronel (1924), who states that in Italy, *Mercurialis annua* frequently develops typical and persistent mycorrhiza, while on the other hand, many perennial herbaceous species are entirely free from it.

In its extended form, the theory of tuberisation outlined in this chapter has been accepted by Constantin (1922), whose conclusions have been summarised as follows: "The association of perennial species of plants with soil fungi has brought about a permanent symbiosis—a condition which does not occur with annual species. Since the perennial character in plants is due to the low temperatures of high altitudes and latitudes, cool climates may be considered as favourable to the establishment of symbiosis. Cultivated Potatoes have lost the mycorrhizal relations of the primitive forms to which tuberisation was due and in order to produce tubers without this relationship they must be grown in cool climates." More than one controversial issue is raised in this passage, but space does not now permit their further discussion.

FIG. 64. Dimorphism and symbiosis. A—D, plants which have suffered infection and formed mycorrhiza; A'—D', plants which have suffered infection, but, owing to rapid digestion of mycelium by the root cells, have not formed mycorrhiza. A, A', *Solanum tuberosum*; B, B', *Orobus tuberosus*; C, C', *Orchis maculata*; D, *Mercurialis perennis*; D', *Mercurialis annua*. *t*, tubercles; *rh*, rhizomes; *b* (fig. C'), main axis of *Orchis maculata*; *b'*, secondary axis developing to a shoot. C and C' after Noël Bernard.
(From Magrou, *Ann. d. Sci. nat.* 1921.)

In conclusion, while recognising the extremely suggestive character of the facts presented by Magrou it appears somewhat doubtful whether sufficient positive evidence has yet been accumulated to serve as a sure foundation for the imposing superstructure raised upon it. In particular, crucial evidence respecting the behaviour of Potato plants with and without infection in aseptic cultures would be welcomed, as also would be an extension of the comparative observations on *Mercurialis* and corresponding types to plants from as many and varied stations as possible. In its present form this stimulating theory of the cause of tuberisation and the origin of the perennial habit should serve as a valuable incentive to further researches in this fascinating field of work.

CHAPTER XI

PHYSIOLOGY AND NUTRITION

The physiological significance of mycorrhiza and the nutrition of mycorrhiza plants—Critical review of evidence derived from researches on forest trees, Orchids and Heaths—Endotrophic infection in other species—Summary and conclusions.

WHAT is the real significance of mycorrhiza and the mycorrhizal habit? It is more than half a century since Frank first asked this question, coining a new name to mark the recognition of root infection by fungi as a normal and regular phenomenon in vascular plants. Of necessity much of the earlier work was descriptive and analytical rather than experimental. The problem bristled with difficulties, some of them very imperfectly understood, and many of the earlier experiments designed to throw light on the biology of the association failed in their object, partly by reason of the prevailing ignorance respecting soil conditions, partly from lack of a suitable technique for isolating the endophytes and establishing their identity. It was not until Bernard applied the methods of the bacteriological laboratory to the study of Orchid mycorrhiza that any real advance was made, and it is now fully recognised by all competent workers at the subject that further knowledge can be gained only by the use of similar methods.

Investigations of the kind indicated are apt to be laborious and they involve the application of a special technique. The isolation of the specific endophytes, their maintenance in pure culture, and the synthesis of fungus and vascular plant under pure culture conditions, all present difficulties. Moreover, evidence of behaviour under rigid experimental control of this kind must be applied with great caution to conditions in nature. Thus it is that the bionomics of mycorrhiza and the nutrition of mycotrophic plants are still, in large measure, the subjects of controversy, although enough is known to correlate them with kindred phenomena of parasitism and the parasitic habit, and to bring them into close touch with the problems presented to the forester and the student of plant ecology in the field.

The present occasion offers a suitable opportunity to review the experimental evidence available and to learn, incidentally, how far it provides support for the more or less speculative theories of nutrition put forward by earlier observers. Two aspects of the problem must receive attention, one relating to green species autotrophic in respect to their carbon nutrition, the other to the non-green forms usually regarded as saprophytic.

Excepting parasites and insectivorous plants, the habit is known to be widespread among terrestrial species, affecting a very large number of families and intensively developed in a certain number of them. Mycorrhiza is found in epiphytes although usually absent from the aerial roots, and, with a single exception, is invariably well developed in non-chlorophyllous "saprophytes," e.g. *Neottia*, *Corallorhiza*, *Monotropa* and its allies. *Wullschlaegelia aphylla*, an Orchid reported by Johow (1889) to be entirely free from fungus infection, is unique in this respect and requires reinvestigation.

Although modern work has not confirmed the view that the ectotrophic and endotrophic habits can be sharply distinguished from one another, it is still convenient to classify mycorrhizas in two main groups, the extreme types of which show marked structural differences, correlated with the distribution and character of infection. In one, a more or less extensive distribution of intracellular mycelium is associated with a variable development of hyphae on the outside of the root; in the other, specially characteristic of trees and shrubs, a sheath of mycelium is formed about the tip and younger part of the root and the mycelium comprising the mantle is continuous with an intercellular net of hyphae in the outer cortex. Intercellular infection, formerly believed to be restricted to the epidermis, has been reported in the cortical tissues of many mycorrhizas of the latter type, so giving rise to the intermediate type of structure named *ectendotrophic*. With few exceptions, the ectotrophic (including the ectendotrophic) habit is limited to woodland soils and its appearance is possibly related to a special environment rather than to any fundamental distinction of a biological kind.

There is no need to labour the point that every possible opinion has been held respecting the physiological relations in mycorrhiza. The nature of these views has been sufficiently indicated in the preceding pages, and it would be tedious to recapitulate the arguments put forward in their support. In the sense intended by de Bary, the relationship is clearly one of *symbiosis*, and the constituents may correctly be described as *symbionts* without any implication that their mutual relation is advantageous or the reverse.

Are the endophytic fungi parasites of a relatively harmless kind or is there a reciprocal relation beneficial to either symbiont or to both?

Two general observations provide evidence of an indirect kind in support of the last-named view. Firstly, it is the young and actively absorbing roots which become mycorrhizas; secondly, there is no cytological evidence of damage to the living constituents of invaded cells. Direct evidence bearing on the enquiry may be derived from three independent sources; namely, the results of experimental investigations on forest trees, Orchids and Heaths respectively.

The mycorrhiza of trees. Melin's work on conifers and deciduous trees has been fully reviewed in an earlier chapter. Certain facts may be regarded as well established. In Sweden a number of Hymenomycetes have been identified as endophytes of Pine and other trees. These forms belong to saprophytic genera, the fruit bodies of which are a constant feature of the woodlands concerned; there is no evidence of parasitism of the ordinary kind nor do they belong to parasitic genera. The ectendotrophic rather than the ectotrophic habit prevails, but such mycelium as becomes intracellular undergoes rapid digestion in the cortical cells. It is likely that some at least of these fungi are obligate symbionts, unable to reach their full development except in association with the roots of their hosts; on the other hand, there is at present no evidence that the early stages of seedling development in any tree is bound up with fungus infection. Pure culture experiments have yielded abundant evidence of specialisation on the part of these root fungi as compared with indifferent soil species. Grown apart from their hosts, they make relatively slow and feeble growth, and show marked sensitiveness to external conditions, e.g. to the hydrogen-ion concentration of the substratum and the character of the food material. Most significant is the observation that normal mycorrhiza is developed only in a suitable rooting medium: its formation is easily inhibited, for example, by a change in the reaction, by unsuitable nutrient, or by the presence of toxic substances such as are formed in humus by heat sterilisation. It is of particular interest to note that, under such adverse conditions, the roots are not subject to a more intense attack. On the contrary, the presence of factors unfavourable to fungus growth inhibits the formation of hyphal complexes in endotrophic mycorrhizas and produces a corresponding reduction of infection in those of ectotrophic type. Under conditions inimical to growth of the seedling the association may lapse into parasitism, while in unfavourable soil conditions pseudomycorrhiza is formed as a consequence of invasion of the roots by soil fungi other than the true endophytes. In short, the formation of mycorrhiza by conifers is shown to be a reciprocal phenomenon conditioned by the physiological states of both symbionts, this in turn being correlated with external conditions of soil and climate. There is good evidence that the fungi profit from the association: they show a marked stimulation from contact with living roots, attributed to the presence of small quantities of exudates, in particular phosphatids, and there can be little doubt that they obtain carbohydrates, especially glucose, from their vascular hosts.

With respect to the absorption of mineral salts other than compounds of nitrogen, Melin carried out a special investigation, making ash estimations of seedlings grown with different combinations and concentrations of salts. He concluded that inorganic salts are taken up as efficiently by infected as by uninfected roots. In view of the relative poverty of free salts in the superficial layers of raw humus as compared with mild humus soils, he was led to the view that absorption of inorganic salts was probably carried on more effectively in the former by mycorrhiza than by uninfected roots—in short to a qualified agreement with the Stahl hypothesis in respect to this matter.

In respect of nitrogenous metabolism the experimental evidence accumulated by Melin is even more striking. In pure culture the endophytes have been found to utilise nucleic acid and like organic compounds much more efficiently than do the seedlings of their hosts. Accordingly, in synthetic cultures supplied with nucleic acid or peptone as sources of nitrogen, the roots of Pine seedlings possessing mycorrhizas do not develop the marked symptoms of nitrogen starvation displayed by those of seedlings grown alone (cf. Fig. 56). On the other hand, there is at present no evidence that any of the known root fungi of trees, whether alone or in association with their hosts, can assimilate free nitrogen.

A critical examination of the experimental evidence confirms the main conclusion reached by Melin, namely, that root infection by fungi possesses a vital significance for trees and plants growing in raw humus soils. In Northern Europe, tree mycorrhiza is typically developed under such conditions, and there is good reason for believing that the soil reaction and the character of the nitrogenous compounds present in the humus are important factors in maintaining mycelium in a physiological condition favourable to its formation. The supply of nitrates is very deficient in these woodland soils. Conifers, and in all probability other trees, can utilise ammonium compounds and possibly to some slight extent, more complex organic compounds of nitrogen, but the last named are much more readily used by the root fungi. Hence, on acid humus soils in which such compounds constitute the chief source of nitrogenous nutrient, plants provided with mycorrhiza are extraordinarily well equipped for competition with other soil organisms. Nutrient material assimilated by the mycelium, if soluble, can enter the root cells directly from the hyphae which invest them, if insoluble, a proportion becomes available by digestion of those hyphae which enter the cells. It is perfectly clear from the experimental evidence now available that the presence of an investing sheath of mycelium about the absorbing region of the roots offers no hindrance to the passage inwards of water and dissolved substances.

In so far as they have been experimentally proved, these conclusions apply only to certain coniferous trees. There can be little reasonable doubt of their ultimate extension to other conifers and also to deciduous trees growing under similar soil conditions and showing a similar mycorrhizal habit. Assuming that the mycorrhizal condition in other trees resembles that in the conifers studied by Melin, it is not possible to doubt the existence of a reciprocal relation conferring benefit upon the trees in certain types of soil by facilitating the intake of nitrogenous food material. While there is good evidence of an advanced degree of specialisation in respect to the symbiotic habit in certain of the endophytes, there is at present none pointing to the existence of an obligate relation or an advanced degree of dependence upon infection on the part of any host tree. In this connection it is perhaps significant that the loss of chlorophyll and the vegetative reduction associated with the saprophytic habit is not known among trees or shrubs, although it is well marked in woody parasites such as the Loranthaceae. The ectotrophic and ectendotrophic types of mycorrhizal structure so common in trees may be related in part to the saprophytic character of the fungi concerned, in part to the abundance of organic residues present in woodland soils.

The conclusions outlined above are supported by the results of Falck's researches on the nutritive significance of mycorrhiza in acid woodland soils ("Rohhumusboden"). (Falck 1923.)

Orchid mycorrhiza. What is the relation between the chlorophyllous Orchids and their root fungi? Does Bernard's conception of it as one of parasitic attack, countered by a mechanism conferring a relative degree of immunity upon the host, cover all the known facts? That it has served to illuminate and explain many aspects of the problem there can be no doubt. Nevertheless, it does not and cannot provide an explanation of the two most characteristic features of the partnership; namely, the obligate character of the connection between infection and seedling development, and the association of specific strains of *Rhizoctonia* with individual Orchid species over wide geographical areas.

Notwithstanding the specialisation incidental to their symbiotic habit, the Orchid fungi appear to retain unimpaired the power of autonomous existence, thereby differing sharply from the root fungi of conifers. They are easily isolated, grow readily in many different media, and produce conidia and sclerotia when grown apart from their hosts. Burgeff reported that the formation of these structures in pure culture was bound up with the concentration of the medium and the amount of metabolate present. The effects of prolonged culture outside their hosts upon the capacity for inducing seedling development are still somewhat obscure and the observations of different workers are at variance.

With respect to nutrition, the salient facts have been described so often and by so many different observers that it is unnecessary to recapitulate them in detail. It can hardly be doubted that the condition observable in living Orchids has been evolved from one following upon invasion of the roots by facultative parasites. In general, the appearance of intracellular mycelium coincides with the disappearance of starch from the invaded cells, and Burgeff showed that in pure culture starch was hydrolysed by diastatic enzymes, the resulting maltose undergoing in certain cases further hydrolysis to glucose. Cane sugar was also removed from the medium, in part inverted, in part directly utilised by the mycelium; glucosides were hydrolysed with removal of the resulting sugars, and cytase and tyrosinase were identified in individual strains. The presence of proteolytic enzymes was likewise reported by the same author. Of nitrogenous nutrients, peptone and salep proved most acceptable, while ammonium compounds were preferred to nitrates. There can be no doubt that the "Eiweisshyphen" that suffer intracellular digestion in the tissues of Orchids contain much protein material, which presumably becomes available to the host cells. This conclusion is supported by the experimental observations of Fuchs and Ziegenspeck (1922) and of Wolff (1925, 1926) on *Neottia*, although, in green Orchids, it is not certain what proportion of it represents material filched from the root cells during the initial phase of fungus activity.

Burgeff found no evidence that any of the Orchid endophytes could utilise atmospheric nitrogen, but H. Wolff (1925) reported nitrogen fixation for the endophyte of *Neottia* in pure culture and has since (1926) confirmed this and extended the report to the root fungi of *Gymnadenia conopsea*, *Orchis maculatus*, *Helleborine palustris* and *H. latifolia*.

The outstanding difficulty in ascribing a beneficent rôle to the root endophytes of green Orchids lies in the fact that, while it is theoretically possible for the mycelium to draw upon food reserves locked up in soil humus, there is in most species but a scanty development of mycelium on the external surface of the roots. On the other hand, it is characteristic of Orchid mycorrhiza that infection does not spread throughout the plant body, but takes place intermittently from without. This, in itself, involves a recurrent "tapping" of the contents of hyphae in direct contact with the soil.

It is possible to regard the obligate relation as a temporary phase related definitely to the developmental stage of growth and dependent, as regards physiological mechanism, upon a rise in the concentration of osmotic substances in the cells of the embryo, but the evidence is still conflicting as to whether or not the mature plants of all Orchids can thrive in the absence of their root fungi. It is of interest to note that different Orchid genera respond differently to seed infection. For example, it was definitely put on record by Bernard that more favourable results were obtained with *Cattleya* seed by using the asymbiotic method of germination, e.g. in his experiments by raising the concentration of salep used in the medium. In the application of pure culture methods of germination to horticultural practice, cases have been recorded in which germination in the neighbourhood of 100 per cent. was secured with *Cattleya* seed, only after more than one strain of the endophyte had been isolated and tested. It is significant that Knudson bases many of his arguments on statements made by Bernard in reference to this genus, members of which often behave in an anomalous manner, as the latter clearly stated.

The successful application of asymbiotic methods to effect germination is not questioned, but they involve the provision of aseptic conditions that do not occur in nature, and are not found to be requisite in horticultural practice when the endophytes are supplied. Moreover, in certain genera, seed development and growth take place more rapidly in fungus-infected cultures than in those protected from infection (cf. Pl. III, Figs. 24, 25), and it is clear from the facts put on record by Clement (1922) that in asymbiotic seed cultures of certain genera the substrata must be very carefully adjusted to the needs of particular species.

In the chlorophyllous Orchids, it is possible to postulate a beneficial effect upon the host in respect to the seedling only, while regarding the mycorrhizal relation in the adult as an incidental evil, combated and held in check by a relatively efficient mechanism of resistance. The occurrence of a number of non-chlorophyllous species in itself raises doubts as to the correctness of this interpretation, and favours the view that the endophytes play an important part in the nutrition of all Orchids. It is tempting to regard the non-chlorophyllous forms as the end terms of a series culminating in complete dependence upon the fungal symbionts. Excluding the possibility of nitrogen fixation by the vascular partners, for which there is no evidence whatever, these species must obtain the whole of their nutrients from the soil, i.e. all the carbonaceous, and the larger part of the nitrogenous materials required must be presented in the form of humus constituents. There are, accordingly, but two possible alternatives in respect to nutrition; either the higher plants can themselves deal directly with the insoluble humus compounds available, or they must obtain the requisite supplies second-hand as a result of the activities of the intracellular mycelium with which the roots are so lavishly supplied. There is no experimental evidence in favour of the former hypothesis, and the provision of such evidence involves the raising of plants free from infection, and their maintenance in aseptic cultures; the alternative view has usually been assumed, and all the arguments applied to its support in the case of green species acquire an added force in that of the non-green forms. The corollary of this tentative conclusion has not received the attention it deserves, namely, that the non-chlorophyllous Orchids are not "saprophytic" but are, in the main, parasites upon their endophytic fungi. It may be observed that all these plants are characteristic members of the vegetations of woodland humus. It follows, therefore, that the benefits conferred upon the vascular hosts in organic soils are common to both ecto- and endotrophic types of mycorrhiza, since *Monotropa* and its allies closely resemble *Neottia* and similar Orchids in respect to the reduction and loss of vegetative features associated directly with the autotrophic habit. The opinion is thus confirmed that the extreme types of structure known as ectotrophic and endotrophic respectively have a similar significance in respect to nutrition, although curiosity is whetted once more as to the details of the physiological mechanism responsible for this result.

In *Gastrodia* there can be little doubt that this is actually the case. For the remainder of this relatively large and extremely interesting group, there is, as yet, no direct experimental proof. The case of *Wullschlaegelia* acquires critical importance in this connection; in view of the improvement in technique since Johow's observations were made, it requires to be reinvestigated.

There is at present a notable lack of comparative data in respect to seed germination in non-chlorophyllous Orchids. In *Neottia* the earlier stages of development were studied by Bernard (1899)[1]. The endophyte is a form included in *Rhizoctonia repens* Bernard, and the details of infection resemble those noted in other genera save in respect to the heavier infection and wider distribution of mycelium observed in *Neottia*. It would be of particular interest to study the behaviour of seeds of *Neottia* and other species lacking chlorophyll under asymbiotic conditions, to ascertain whether the stimulus normally exerted by the endophytes can be replaced by the addition of organic substances to the substratum, and if so, estimate the capacity of the resulting seedlings to assimilate organic nutrients in the absence of mycelium.

The remarkable case of *Gastrodia* (Kusano, 1911) also challenges attention as one in which an association with a fungus of the *Rhizoctonia* type apparently does not occur and for which there is at present no information respecting seed germination or the earlier stages of seedling development.

It was in Orchids that infestation of the roots by fungus mycelium was first clearly recognised. In spite of the relatively exact knowledge that has been acquired, there remains still much that is puzzling in the relation of fungus and host in this group of plants. In the evolutionary sense, it must have originated in parasitic attack on the part of certain soil fungi adequately combated and held in check by the victims. This view assumes the existence of relatively benign strains of the attacking fungus, or of individual Orchid plants possessing relatively high resistance to parasitic invasion of the roots. In species frequenting humus soils and in epiphytes, root infection may have resulted in the "tapping" of organic residues otherwise inaccessible, and undoubtedly this would be facilitated by the

[1] "J'ai eu la bonne fortune de trouver, dans la Nature, des milliers de graines en germination d'une Orchidée indigène (*Neottia Nidus-avis*)." (Bernard 1902.)

intermittent character of infection. In those Orchid species specially associated with chalky and limestone soils, it is unlikely that this source could yield food supplies of any significance.

The obligate relation is clearly secondary: it is nutritive in origin and probably affects only the early stages of development; confirmatory evidence is required that it persists in the adult plant.

The mycorrhiza of Heaths. Turning to Ericaceae, another group for which critical results are available, it may be noted that while an obligate association with specific endophytes is common to Orchids and Heaths, the structural features of the mycorrhiza in plants belonging to the latter group resemble those in conifers rather than in Orchids. The incidence of infection in the tissues is more superficial, and there is a more profuse development of mycelium on the surface of the roots. In *Calluna*, there is a tendency for hyphae to accumulate about the root tips, and in other genera, e.g. *Vaccinium* and *Arbutus*, the younger part of the roots may be invested by a complete sheath of mycelium—the tuberous mycorrhiza of *Arbutus unedo*, for example, is typically ectendotrophic in structure.

Now *Calluna* and many of its allies are characteristic and abundant members of the vegetation of humus soils—moorland, heath, and woodland—a notable feature of such soils being a deficiency of mineral salts, especially nitrates. Under field conditions, the intensity of infection in *Calluna* varies. As observed by the writer, mycorrhiza is profusely developed in soils with abundant humus: on peaty heaths and moorlands, and in cultivated garden soils. It is more sparsely formed in dry sandy situations, where, at certain times of the year, its roots although infected may appear to be free from mycelium. In experimental cultures this immunity from infection may be even more striking. Thus, it is now believed by the writer that *functional* mycorrhiza is not formed in aseptic cultures, whether sand or agar-agar, supplied with inorganic nutrient solutions, or in peat sterilised by heat, the latter being temporarily extremely toxic as a rooting medium. Cytological observations, as yet incomplete (see p. 213), indicate that a marked change in the metabolic processes leading to the appearance of fats in the root cells, coincides with suppression or reduction of the usual hyphal complexes. The significance of such changes in relation to nutrition has not yet been investigated.

These fluctuations are similar in kind to those observed in conifers, and they appear to be correlated with corresponding differences in the rooting medium, i.e. in nature, variation in the

14-2

amount and character of the organic constituents; under experimental conditions, the absence of organic compounds or the presence of toxic substances such as those produced by heat sterilisation.

Hence, in *Calluna* as in conifers, the production of healthy and functional mycorrhiza seems to be directly related to the presence of factors favourable to the growth of the mycorrhizal fungi, in effect, to soil conditions not readily reproduced under "pure culture" conditions.

A case for exchange of nutrient material with a balance of profit on the side of the vascular plant has been set out, and the opinion expressed that in *Calluna*, the endophyte operates as an internal factor of a special kind directly related to the metabolism of the root cells in certain kinds of soil. Evidence has been supplied that the formation of mycorrhiza is "a reciprocal phenomenon, conditioned not only by the activity of the fungus, but by the reaction of the root cells and the nature of the rooting medium" (Rayner, 1925).

In view of the reciprocity observed both in conifers and in *Calluna* between the constitution of the rooting medium and the character of infection, it seems probable that Melin's conclusions respecting the beneficial effects resulting from the presence of the mycorrhiza in acid humus may be extended to the more specialised case of *Calluna*. Under such conditions, the mycorrhiza of this species, and doubtless of other ericoids, probably functions in a similar way to that of trees, conferring on the host plant the power of drawing upon the organic food reserves locked up in humus. In Ericaceae, the relation is more specialised than in trees. In each species infection is restricted to a single endophyte showing a relatively high degree of specificity, and there is positive evidence for believing that the capacity for utilising organic compounds of nitrogen is supplemented by nitrogen fixation on the part of the endophytes (Rayner, 1922, 1925).

On this view the mycorrhizal habit in *Calluna* and its allies may be regarded as an attempt—as successful in its way, although attended by greater risks to the host, as that manifested in the tubercles of legumes and other nodule-forming species—to solve the problem of obtaining the requisite supplies of nitrogen on soils deficient in nitrates. Like certain conifers and other trees, *Calluna* and its allies are not strictly autotrophic in respect to their nitrogen metabolism, and they are singularly well equipped for successful competition in the struggle to obtain the requisite nitrogenous food materials, whether in sandy soils, poor in organic constituents, or in acid humus soils deficient in nitrates. They have "solved the problem of growth upon the poorest and most unpromising soils, but they have solved it at the price of their independence" (Rayner, 1915).

Although infection by the appropriate endophyte is an obligate condition for seedling development in *Calluna* and other ericaceous species, there is no proof that the formation of typical mycorrhiza is essential to the mature plants, albeit there is good evidence that it renders available sources of nitrogen otherwise inaccessible. The exact nature of the benefit so gained may fluctuate with the edaphic character of the habitat; for example, nitrogen fixation may be in abeyance when organic compounds of nitrogen are abundant.

With regard to the exchange of organised food materials in mycorrhiza plants generally, McLennan has provided evidence that the root fungus of *Lolium* forms fats; in the later stage of growth the "sporangioles" become gorged with oily material transferred from the adjoining hyphae. The fat is ultimately discharged into the cells, where, presumably, it is hydrolysed, and whence it is translocated during the fruiting season of the host (Pl. VII, Figs. 61, 62)

The cytology of the infected tissues in *Lolium* and *Gastrodia* has been compared. McLennan interprets the physiological relation in both species as a metabolic exchange "from the fungus to the higher plant, with the result that the latter obtains a supply of fat or oil" (McLennan, 1926).

The author lays special emphasis on the exchange of carbonaceous rather than nitrogenous food materials, as deduced from cytological features in *Gastrodia* and *Lolium*. But it must be noted that the former plant is likewise dependent upon the endophyte for the whole of its nutrient material, and there appears to be no reason to assume that in *Lolium* the translocations of fatty and nitrogenous food materials are alternative or mutually antagonistic processes. More reasonable is the view that the exact nature of the exchange in each case is intimately bound up with the metabolism of the particular host plant, with the stage in development, and possibly, in some cases, indirectly with edaphic peculiarities. A similar transference of fatty material from mycelium to root cells, followed by removal of the stainable material, takes place in the root cells of *Calluna* during digestion of the mycelial complex. It is believed that the fat metabolism of the latter plant, and presumably that of other ericaceous species, may have an important bearing on their nutritive relations with the endophytes, and also with their reaction to calcareous soils, and that a clearer understanding of it may throw further light on the significance of the calcifuge habit in Ericaceae. It is hoped that work now in progress may render it possible to relate the two sets of phenomena more precisely.

Although members of the Monotropoideae have long attracted attention, no experimental data are yet available respecting germination, seedling development, or the real significance of the saprophytic habit in members of this group of Ericales. In view of the obligate relation in Heather and kindred species, it is tempting to regard them as the end terms of a series culminating in complete dependence upon their fungus associates. Whether or not this is actually the case, an experimental study of these interesting plants is long overdue. Unsuccessful attempts to germinate seed of *Monotropa hypopitys* have been made by more than one observer including the writer, and it is already certain that germination cannot be effected by the ordinary asymbiotic methods.

There is little experimental evidence bearing on the exact nature of the biological relation in mycorrhiza plants other than those included in the specialised groups just considered. The uniform type of infection, i.e. by mycelium bearing vesicles and arbuscules, and the prevalence of intracellular digestion favour the conclusions that the endophytes of many vascular plants belong to a single group, and that the physiology of the relation does not differ fundamentally from that in the cases studied experimentally. No satisfactory evidence exists that an endophyte of this type has yet been isolated, in itself an indication of specialisation to the symbiotic habit: nor is anything known about the bionomics of "double infection" as reported by Peyronel for many mycorrhizal plants that show primary infection by mycelium bearing vesicles and arbuscules. Save in respect to its mode of entry, the latter appears to behave as a saprophyte rather than as a parasite. Starch and possibly other metabolites are removed from the cells, although any benefit accruing to the fungus must be evanescent in view of the extensive digestion of mycelium which eventually takes place. It is possible that a beneficial effect upon the host may be related directly to soil conditions; for example, in soils rich in natural humus or heavily manured, the mycelium may bring in material derived from organic residues and so restore more than was originally removed from the root cells.

The condition in *Lolium*—more particularly in view of the possibility that the hyphae in root and shoot respectively belong to the same fungus—together with the appearance of the "saprophytic" habit in certain families, e.g. Lycopodiaceae, Ophioglossaceae and Gentianaceae, suggests that a relatively advanced condition has already been evolved in certain groups. The physiology of the symbiotic relation doubtless varies from species to species—possibly from individual to individual. Only by experiments with infected and uninfected plants can final conclusions be reached as to the effects of infestation upon the hosts. In the relatively favourable conditions inseparable from pure culture experiments, it is not impossible that plants normally subject to mycorrhiza formation would thrive better protected from infection. Great caution is therefore required in applying results deduced from experimental cultures to plants growing in nature. Whereas under "sheltered" conditions, mycorrhiza formation may be incidental to seedling infection and of little significance in nutrition, it may become of critical importance to plants exposed to the full brunt of competition in the field.

In its evolutionary aspect the relation now existing in Orchids and Heaths must have originated in parasitism on the part of certain soil fungi. Assuming the existence of strains capable of invading living roots but relatively benign in their action and likewise the presence of individual hosts with a relatively high resistance to parasitic attack, it is not difficult to reconstruct the stages by which a symbiotic association was initiated. The endophytic habit brought with it certain changes in the metabolism of both partners, clues to the nature of which may be sought in the response shown by Orchid embryos to an increased concentration of sugar in the substratum, and in the still unknown mechanisms by which seedling development is activated in *Calluna* and other ericoids.

On any hypothesis, the evolution of an obligate relation with a parasitic or facultatively parasitic fungus is difficult to explain, and raises question of profound interest to students of heredity. Both in the Orchids and in *Calluna* it is associated now with an early stage of seedling development and in neither case is there satisfactory proof that it persists in the adult. From the reaction of Orchid embryos towards increased content of sugar or other organic nutrients in the substratum, it may be surmised that the association is intimately bound up with metabolic changes induced by the presence of the endophyte in the tissues.

The mycorrhizal habit in trees appears to have had a somewhat different origin and to have evolved along rather different lines.

Originating in the attraction exerted on saprophytic soil fungi by the presence of exudates—especially phosphatids—about the growing roots, it seems possible that the ectotrophic habit resulted from the feebly parasitic character of the fungi present, and that this structure in its purest form represents an early stage in the evolution of tree mycorrhiza. With the symbiotic habit thus established came in time a more efficient mechanism for penetrating the root tissues, the extension of intercellular infection to the more deeply placed root tissues and the appearance of the ectendotrophic type of structure. There is in trees no evidence of an obligate relation with seedling development and the formation of typical mycorrhiza appears to be bound up with the utilisation of the organic residues in humus. Producing comparatively small effects under the sheltered conditions inseparable from experimental culture, the presence of active mycorrhiza may have become a critical factor in nature, determining the survival value of the host trees in the competitive struggle for the available nutrients in humus and woodlands soils. Viewed from this angle the comparative observations of Möller and others, on the condition of roots in respect to infection in soils of differing character, acquire an added significance and find their place in the theory of nutrition put forward recently by Melin.

It is well to recall the fact that in the plant world, the severity of the struggle for existence not uncommonly centres about the competition for suitable compounds of nitrogen. Many of the most striking "adaptations" known are directly related to it, e.g. the insectivorous habit, and the formation of root nodules as a reaction to invasion of the tissues by nitrogen-fixing bacteria. Even the rôle of vascular plants in general as "nitrate organisms" may be so regarded, forming, as it does, an indispensable link in that remarkable cycle of chemical changes by which the "circulation of nitrogen" is secured and maintained in nature. On *a priori* grounds, it is not unreasonable to believe that the intimate association with fungus mycelium, so common in all groups of vascular plants and also in the thalloid members of Bryophyta, is but another manifestation of the urgency of this nitrogen problem among plants. In at least two of the three groups of mycorrhizal plants for which experimental data are now available, namely, forest trees and members of Ericaceae, modern work has yielded corroborative evidence that this is actually the case.

For Orchids, the evidence in support of this general hypothesis is less convincing, although there can be no doubt that nitrogenous substances are present in rich abundance in the "Eiweisshyphen" which constitute so conspicuous a feature of the intracellular mycelium within the root tissues.

How the root fungi can derive any but a very temporary benefit from the association is not evident. In the case of tree mycorrhiza, Melin's researches lead him to postulate the existence of a large group of fungi belonging to the Hymenomycetes which cannot rightly be described as either parasites or saprophytes. Physiologically, they stand in closer relation to parasites and it is suggested they should be known as *Symbiophiles*.

The endophytes of Orchids and Heaths have definite affinities with parasitic forms and the primary relation with their vascular hosts must undoubtedly have been one of "attack and defence" as postulated by Bernard. From this has been evolved a condition of equilibrium, relatively stable in character although easily disturbed by changes in the environment, with the balance of profit—certainly in *Calluna* and allied species and probably also in Orchids—definitely on the side of the vascular partners.

Specialisation has been accompanied by physiological changes affecting the metabolism of both partners, notably in regard to the appearance of an obligate relation of a temporary kind in the early stages of seedling development. The origin and precise significance of the latter are at present obscure; it may be possible to relate it, on the one hand with metabolic changes associated with the endophytic habit, on the other with the saprophytic mode of nutrition normal to the embryo stages of seedling development.

What bearing have these conclusions on the theories of nutrition put forward by the earlier workers? Allusion has been made to the views expressed by Hartig and others respecting parasitism on the part of the root fungi of trees. Melin's conclusions, on the contrary, support the theory of beneficial symbiosis first put forward and maintained by Frank and his school. Is it possible to reconcile these divergent opinions? Is it, for instance, worth enquiring whether there is any evidence that the equilibrium implied in the existence of a symbiosis beneficial to the trees is ever disturbed in nature? Under experimental conditions such disturbance has been noted with resulting injury to the hosts. It has not been actually observed in the field, but Melin has expressed the opinion that overthrow of the "balance of power" ordinarily maintained, may easily occur during the earlier stages of growth. If seedlings are feeble or the mycelium too vigorous, there is a tendency for the root fungi to become parasitic; if the young trees are healthy and the mycelium well nourished, equilibrium is established and maintained and the relation may fairly be described as *mutualism*.

On the whole it may be said that the views so strongly held by Frank, Stahl and their followers have survived the test of experimental enquiry. The arguments employed are not always acceptable in the light of modern research, but, in the opinion of the writer, there can be no doubt that recent investigations by means of pure cultures have tended to support the view that the possession of mycorrhiza is frequently of benefit to the vascular hosts, the nature and extent of such benefit depending upon the physical conditions of the environment and the physiology of the association in individual cases.

BIBLIOGRAPHY

ACTON, E. H. (1889). The assimilation of carbon by green plants from certain organic compounds. *Proc. R. Soc. London*, **46**, B, 1889, p. 118. (Abstract, *Bot. Centralbl.* **44**, 1890, p. 224.)

ARRHENIUS, O. (1920). *Ökologische Studien in den Stockholmer Schären.* Inaug. Dissert. Stockholm, 1920.

—— (1922). *Bodenreaktion und Pflanzenleben mit spezieller Berücksichtigung des Kalkbedarfs für die Pflanzenproduktion.* Leipzig, 1922.

ATKINSON, G. F. (1892). The genus *Frankia* in the United States. *Bull. Torrey Bot. Club*, **19**, 1892, p. 171.

BACCARINI, P. (1903). Sopra i caratteri di qualche *Endogone*. *Nuova Giorn. Bot. Ital.* n. ser., **10**, 1903, p. 79.

BARTHEL, C. (1923). *A review of the present problems and methods of agricultural bacteriology.* Works published by Knut and Alice Wallenberg Foundation, 1. Stockholm, 1923.

BARY, A. de (1858). Sur la germination des Lycopodées. *Ann. d. Sc. nat. Bot.* **4**, 1858, p. 30.

—— (1879). *Die Erscheinung der Symbiose.* Vortrag gehalten auf der Versammlung deutscher Naturforscher und Aerzte zu Cassel (1878). Strasbourg, 1879, 8°.

BAUMANN, A. (1886). *Ueber die Bestimmung des im Boden enthaltenden Ammoniakstickstoffs u. ueber die Menge des assimilirbaren Stickstoffs im unbearbeiteten Boden.* Habilitationschrift. 1886.

BEAL, W. J. (1890). Root-galls in *Ceanothus americanus*. *Bot. Gazette*, **15**, 1890, p. 232.

BEAUVERIE, J. (1902). Étude d'une Hépatique à thalle habité par un champignon filamenteux. *C.R. Acad. des Sc.* **134**, 1902, p. 616.

BECK, G. (1880). Einige Bemerkungen über den Vorkeim von *Lycopodium*. *Oest. Bot. Zeit.* **30**, 1880, p. 341.

BEIJERINCK, M. W. (1888). Die Bakterien der papilionaceen Knöllchen. *Bot. Zeitung*, **46**, 1888, cols. 725, 741, 757, 781, 797.

—— (1889). *Handelungen van het tweede nederlandsch Natur-en Geneeshundig Congres*, 1889, p. 108.

—— (1890). Künstliche Infection von *Vicia Faba* mit *Bacillus radicicola*. *Bot. Zeitung*, **52**, 1890; *Bot. Centralbl.* **45–46**, 1891, p. 247.

—— (1901). Ueber oligonitrophile Mikroben. *Centralbl. f. Bakt.* II, **7**, 1901, p. 561.

—— (1904). De invloed der mikroben op de vruchbaarheit van den groud en op den groei der hoogere planten. *Landbouwkundig Tydschrift*, 1904.

BERNARD, NOËL (1899). Sur la germination du *Neottia Nidus-avis*. *C. R. Acad. des Sc.* **128**, 1899, p. 1253.

—— (1900). Sur quelques germinations difficiles. *Rev. gén. de Bot.* **12**, 1900, p. 108.

—— (1902 *a*). Infection et tubérisation chez les végétaux. *Rev. gén. de Bot.* **13**, 1902, p. 8.

—— (1902 *b*). Études sur la tubérisation. *Rev. gén. de Bot.* **14**, 1902, pp. 5, 58, 101, 170, 219, 269.

—— (1902 *c*). Conditions physiques de la tubérisation chez les végétaux. *C. R. Acad. des Sc.* **135**, 1902, p. 706.

—— (1902 *d*). Mécanismes physiques d'actions parasitaires. *Bull. de la Soc. linnéenne de Normandie*, 5me sér. **6**, 1902, p. 127.

BERNARD, NOËL (1903). La germination des Orchidées. *C. R. Acad. des Sc.* **137**, 1903, p. 483.
—— (1904 *a*). Recherches expérimentales sur les Orchidées. *Rev. gén. de Bot.* **16**, 1904 p. 405.
—— (1904 *b*). Le champignon endophyte des Orchidées. *C. R. Acad. des Sc.* **138**, 1904, p. 828.
—— (1905). Nouvelles espèces d'endophytes d'Orchidées. *C. R. Acad. des Sc.* **140**, 1905, p. 1272.
—— (1906 *a*). Symbiose d'Orchidées et de divers champignons endophytes. *C. R. Acad. des Sc.* **142**, 1906, p. 52.
—— (1906 *b*). Les champignons des Orchidées, leur utilisation et leur rôle. *Orchis*, April 1906.
—— (1906 *c*). On the germination of Orchids. *Report of the 3rd International Conference on Genetics, London*, 1906.
—— (1908). *La culture des Orchidées dans ses rapports avec la symbiose*. Pub. by Soc. roy. d'Agric. et de Bot. de Gand, Ghent, 1908.
—— (1909 *a*). L'évolution dans la symbiose. L'Orchidées et leur champignons commensaux. *Ann. d. Sc. nat. Bot.* 9me sér. **9**, 1909, p. 1.
—— (1909 *b*). Remarques sur l'immunité chez les plantes. *Bull. de l'Instit. Pasteur*, **7**, 1909, p. 369.
—— (1909 *c*). L'origine de la pomme de terre. *Bull. Soc. acad. d'Agric., etc., de Poitiers*, déc. 1909.
—— (1911 *a*). Les mycorhizes des *Solanum*. *Ann. d. Sc. nat. Bot.* 9me sér. **14**, 1911, p. 235.
—— (1911 *b*). Sur la fonction fungicide des bulbes d'Ophrydées. *Ann. d. Sc. nat. Bot.* 9me **sér. 14**, 1911, p. 223.
BERNARD, MME NOËL et MAGROU, J. (1911). Sur les mycorhizes des Pommes de terre sauvages. *Ann. d. Sc. nat. Bot.* 9me sér. **14**, 1911, p. 252.
BERNATSKY, J. (1899). Beiträge zur Kenntniss der endotrophen Mycorrhizen. *Természetragzi Filizetek*, 1899, Budapest.
—— (1900). Ueber Mycorrhizengebilde. *Természetragzi Filizetek*, **23**, 1900, Budapest.
BERTHELOT et ANDRÉ, G. (1892). Sur l'oxydation spontanée de l'acide humique et de la terre végétale. *C. R. Acad. d. Sc.* **114**, 1892, p. 41.
BEWLEY, F. and HUTCHINSON, H. B. (1920). Changes through which the nodule organism (*P. radicicola*) passes under cultural conditions. *Journ. of Agric. Sc.* **10**, 1920, p. 144.
BLACKMAN, V. H. (1903). Some recent observations on mycorrhiza. *New Phytologist*, **2**, 1903, p. 23.
BORNEBUSCH, C. H. (1914). Studier over Rødællens Livskrav og dens Optræden i Danmark. [Studies on the ecology of *Alnus glutinosa* and its occurrence in Denmark.] *København Tids. Skov.* **26**, B, 1914, p. 28.
BOTTOMLEY, W. B. (1907). The cross-inoculation of the nodule-forming bacteria from leguminous and non-leguminous plants. *Report Brit. Ass. f. Adv. of Sc.* Leicester, 1907, p. 693.
—— (1907). The structure of root tubercles in legumes and other plants. *Proc. Roy. Soc. London*, **81**, B, 1907, p. 287.
—— (1912). The root nodules of *Myrica Gale*. *Ann. of Botany*, **26**, 1912, p. 111.
—— (1915). The root nodules of *Ceanothus americanus*. *Ann. of Botany*, **29**, 1915, p. 605.
BOUDIER, É. (1876). Du parasitisme probable de quelques espèces du genre *Elaphomyces* et de la recherche de ces tubéracés. *Bull. de la Soc. bot. de France*, **23**, 1876, p. 115.
BOULET, V. (1910). Sur les mycorhizes endotrophes de quelques arbres fruitiers. *C. R. Acad. d. Sc.* **150**, 1910, p. 1190.
BOWER, F. O. (1904). *Ophioglossum simplex*. *Ann. of Botany*, **18**, 1904, p. 205.
—— (1908). *The Origin of a land flora*. London; Macmillan, 1908. p. 240

BOYER, G. (1915–16). Sur l'existence et les principaux caractères du mycélium des champignons qui paraissent en être dépourvus et en particulier de celui des tubéracées. *Actes Soc. linn. Bordeaux*, **69**, 1915–16, p. 94; *Bot. Abs.* **6**, 190, 1920–21, p. 26.

BRANDT, E. (1849). Nonnulla de parasitis quibusdam phanerogamicis observata. *Linnaea*, **22**, 1849, p. 81.

BRÉAL, E. (1894). Alimentation des végétaux par l'humus et les matières organiques. *Ann.* [illegible], **20**, 1894, p. 353.

BREFELD, O. (1908). [illegible] *aus dem Gesamtgebiete der Mykologie.* 14. *Die Kultur der Pilze.* Münster, 1908.

BRENCHLEY, W. E. and THORNTON, H. G. (1925). The relation between the development, structure, and function of the nodules on *Vicia faba*, as influenced by the presence or absence of boron in the nutrient medium. *Proc. Roy. Soc. London*, **98**, B, 1925, p. 373.

BRENNER, W. (1924). Über die Reaktion finnländischer Böden. *Kom. Finl. Agrogeol. Medd. Helsingfors*, **19**, 1924, p. 32.

BRITTON, E. G. (1911). Fungi on Mosses. *The Bryologist*, **14**, 1911, p. 103.

BRUCHMANN, H. (1874). Ueber Anlage u. Wachstum der Wurzeln von *Lycopodium* u. *Isoëtes*. *Jenaische Zeitschrift f. Naturwissenschaft*, **8**, 1874, p. 522.

—— (1885). Das Prothallium von *Lycopodium*. *Bot. Centralbl.* **21–22**, 1885, pp. 23, 309.

—— (1897). *Untersuchungen über* Selaginella spinulosa. Gotha: Perthes, 1897.

—— (1898). *Ueber die Prothallien u. die Keimpflanzen mehrerer Europaischen* Lycopodien. Gotha, 1898. p. 108.

—— (1904). Ueber das Prothallium u. die Keimpflanze von *Ophioglossum vulgatum* L. *Bot. Zeitung*, **62**, 1904, p. 227.

—— (1906). Ueber das Prothallium u. die Sporenpflanze von *Botrychium lunaria*. *Flora*, **96**, 1906, p. 203.

BRUNCHORST, J. (1885). Ueber die Knöllchen an den Legumenosenwurzeln. *Ber. d. d. bot. Gesell.* **3**, 1885, p. 241.

—— (1886). Ueber einige Wurzelanschwellungen, besonders diejenigen von *Alnus* und den Eleagnaceen. *Unters. a. d. bot. Inst. z. Tübingen*, **2**, 1886, p. 151.

—— (1887). Die Structure der Inhaltskörper in den Zellen einiger Wurzelanschwellungen. *Bergens Museums Aarsberetning*, 1887, p. 235.

BRUNS, E. (1894). Beitrag zur Kenntniss der Gattung *Polysaccum*. *Flora*, **78**, 1894, p. 67.

BUCHOLTZ, F. W. (1911). Über die Befruchtung von *Endogone lactiflua*. *Ann. Mycologici*, **9**, 1911, p. 328.

—— (1912). Beiträge zur Kenntniss der Gattung *Endogone* Link. *Beih. z. Bot. Centralbl.* **29**, 1912, p. 147.

BULTEL, G. (1920). Note sur la germination des Graines d'Orchidées à l'aide du champignon endophyte. *Journ. Soc. hort. France*, 4me sér. **21**, 1920, p. 434.

—— (1925). Germinations aseptiques d'Orchidées. *Rev. Hort.* **96**, 1925, pp. 268, 291; **117**, pp. 318, 334, 359.

BURGEFF, H. (1909). *Die Wurzelpilze der Orchideen*, Jena, 1909.

—— (1911). *Die Anzucht tropischer Orchideen aus Samen.* Jena: G. Fischer. 1911.

—— (1913). Symbiose. *Handw. d. Bioch. d. Naturw.* **9**, Jena, 1913.

BÜSGEN, M. (1901). Einiges Untersuchungen über Gestalt u. Wachstumweise der Baumwurzeln. *Allg. Forst- u. Jagdzeit.* **77**, 1901.

BUSICH, E. (1913). Die endotrophe Mykorhiza der Asclepiadaceae. *Verhandl. der K. K. zool.-bot. Gesellschaft, Wien*, **63**, 1913, p. 240.

CAMPBELL, D. H. (1895). *The structure and development of the Mosses and Ferns.* London and New York. 3rd ed. 1918, pp. 238, 239, 270, 624, 625, 635.

—— (1907). Studies in the Ophioglossaceae. *Ann. du Jard. bot. de Buitenzorg*, 2me sér. **6**, 1907, p. 138.

CAMPBELL, D. H. (1908). Symbiosis in fern prothalli. *Amer. Naturalist*, **42**, 1908, p. 154.
—— (1911). *The Eusporangiatae. Comparative Anatomy of the Ophioglossaceae and Marattiaceae.* Carnegie Inst. of Washington, No. **141**, 1911, p. 224.
—— (1911). The structure and affinities of *Macroglossum Alidae*. *Ann. of Botany*, **28**, 1914, p. 651.
—— (1921). The gametophyte and embryo of *Botrychium obliquum* Mühl. *Ann. of Botany*, **35**, 1921, p. 141.
CASPARY, R. Über die Gefässbundeln der Pflanzen. *Monatsber. d. k. preuss. Akad. d. Wissensch. Berlin*, 1862, p. 448.
CAVARA (1896). *Ipertrophe ed anomalie nucleari in sequito a parasitismo vegetale.* Inst. R. de Pavie, 1896. (See also *Rev. Mycol.* **19**, 1897, p. 94: *Riv. Patol veg.* **5**, 1897, p. 238.)
CAVERS, F. (1903). On saprophytism and mycorhiza in Hepaticae. *New Phytologist*, **2**, 1903, p. 30.
—— (1904). On the structure and biology of *Fegatella conica*. *Ann. of Botany*, **18**, 1904, p. 87.
CEILLIER, R. (1912). *Recherches sur les facteurs de la répartition et sur la rôle des mycorhizes.* Thèse: Paris, 1912. (See *Trans. Mycol. Soc.* **8**, 1922, p. 51.)
CHAMBERLAIN, C. J. (1917). Prothallia and sporelings of three New Zealand species of *Lycopodium*. *Bot. Gazette*, **63**, 1917, p. 51.
CHAN, AALI BEDR. T. (1923). Über die Mycorrhiza der Buche. *Allg. Forst- u. Jagdzeit.* **99**, 1923, p. 25. (See also *Review of App. Mycol.* **2**, 1923–24, p. 463.)
CHARLES, G. M. (1911). The anatomy of the sporeling of *Marattia alata*. *Bot. Gazette*, **51**, 1911, p. 81.
CHATIN, A. (1856). *Anatomie comparée des végétaux. Plantes aquatiques et parasitières.* Paris, 1856–65.
CHAUDHURI, H. und RAJARAM (1925). Ein Fall von wahrscheinlichen Symbiose eines Pilzes mit *Marchantia nepalensis*. *Flora*, **22**, 1925, p. 176.
CHIOVENDA, E. (1920). Nuova localita italiana per il *Myriostoma coliliforme*. *Nuova Giorn. Bot. Ital.* n. ser. **27**, 1920, p. 7. (See also *Bot. Abs.* 7, 364, p. 53.)
CHODAT, R. et LENDNER, A. (1896). Sur les mycorhizes du *Listera cordata*. *Bull. de l'Herbier Boissier*, **4**, 1896, p. 265. (See also *Revue Mycol.* **20**, 1898, p. 10.)
CHRISTENSEN, H. R. (1906). Ueber das Vorkommen u. die Verbreitung des *Azotobacter chroococcum* in verscheiden Böden. *Centralbl. f. Bakt.* **17**, 1906, pp. 109, 161, 378.
CHRISTOPH, H. (1921). Untersuchungen über die mykotrophen Verhältnisse der Ericales u. die Keimung von Pirolaceae. *Beih. z. Bot. Centralbl.* **38**, 1921, p. 115.
CLEMENT, E. (1924). Germination of *Odontoglossum* and other seed without fungal aid. *Orchid Review*, **32**, 1924, p. 232.
—— (1924). The non-symbiotic germination of orchid seeds. *Orchid Review*, **32**, 1924, p. 359.
COCKAYNE, L. (1921). *The vegetation of New Zealand.* Leipzig, 1921.
CONDAMY, A. (1876). *Étude sur l'histoire naturelle de la Truffe.* Angoulême, 1876. (Review in *Bull. de la Soc. bot. de France*, **23**, 1876, p. 119.)
CONSTANTIN, J. (1922). Sur l'hérédité acquise. *C. R. Acad. d. Sc.* **174**, 1922, p. 1659. (See also *Bot. Abs.* **13**, 1554, 1924, p. 234.)
—— (1924 *a*). Remarques sur les relations des arbres avec les champignons souterrains. *C. R. Acad. d. Sc.* **178**, 1924, p. 158.
—— (1924 *b*). Les Mycorhizes et la pathologie végétale. *Rev. Bot. appliquée*, **4**, 1924, p. 497.
—— (1925). Remarques sur les cultures asymbiotiques. *Rev. Path. Vég. et Ent. Agric.* **12**, 1925, p. 191.
CONSTANTIN, J. et DUFOUR, L. (1920). Sur la biologie du *Goodyera repens*. *Rev. Gén. de Bot.* **32**, 1920, p. 529.

CONSTANTIN, J. et DUFOUR, L. (1921). Recherches sur la biologie du *Monotropa*. *C. R. Acad. d. Sc.* **173**, 1921, p. 957.

CONSTANTIN, J. et MAGROU, J. (1922). Applications industrielles d'une grande découverte française. *Ann. d. Sc. nat. Bot.* sér 10, **4**, 1922. Actualités biologiques, p. i.

CONSTANTIN, J. et MATRUCHOT, L. (1898). Essai de culture du *Tricholoma nudum* *C. R. Acad. d. Sc.* **126**, 1898, p. 853.

COOKE, M. C. (1889). Some exotic fungi. *Grevillea*, **17**, 1889, p. 76.

CORDEMOY, H. J. DE (1904). Sur une fonction spéciale des mycorhizes des racines latérales de la Vanille. *C. R. Acad. d. Sc.* **138**, 1904, p. 391.

CORTESI, F. (1912). Sulle microrize endotrofiche con particolare riguardo a quelle delle Orchidaceae. *Atti della Società ital. Rome*, 1912, p. 860.

COULTER, J. M., BARNES, C. R. and COWLES, H. C. (1911). *Textbook of Botany*, 1911. pp. 794, 799.

COUNCILMAN, W. T. (1923). The root-system of *Epigaea repens* and its relation to the fungi of the humus. *Proc. Nat. Acad. Sc.* **9**, 1923, p. 279. (See also *Bot. Gaz.* **76**, 1923, p. 428.)

—— (1924). The relation between the roots of plants and fungi. *Proc. Soc. Exp. Biol. and Med.* **21**, 1924, p. 31.

COVILLE, F. V. (1910). Experiments in blueberry culture. *U.S. Dept. of Agric. Bur. of Plt. Industry Bull.* **193**, 1910.

—— (1916). Directions for blueberry culture. *U.S. Dept. of Agric. Bur. of Plt. Industry*, Bull. **334**, 1916.

—— (1921). Directions for blueberry culture. *U.S. Dept. of Agric. Bur. of Plt. Industry, Bull.* **974**, 1921.

CZAPEK, F. (1889). Zur Chemie der Zellmembranen bei den Laub- u. Lebermoosen. *Flora*, **86**, 1889, p. 361.

—— (1896). Zur Lehre von den Wurzelauscheidungen. *Jahrb. f. Wiss. bot.* **29**, 1896, p. 388.

—— (1920). Die organische Ernährung bei höheren grünen Pflanzen. *Naturwissenschaften*, **8**, 1920, p. 226. (See also *Bot. Abs.* **7**, 1307, 1921, p. 194.)

DANGEARD, P. A. et ARMAND, L. (1896). Observations de biologie cellulaire. *Le Botaniste*, 5me sér. **5**, 1896–97, p. 289.

DARWIN, CHAS. (1882). *The various contrivances by which Orchids are fertilised by insects.* London: Murray, 1882.

DEGENER, O. (1924). Four new stations of *Lycopodium* prothallia. *Bot. Gazette*, **77**, 1924, p. 89.

DÉHÉRAIN, T. T. (1889). Recherches sur l'épuisement des terres arables par la culture sans engrais. *Ann. agronomiques*, **15**, 1889, p. 481.

DEMETER, K. (1923). Ueber Plasmoptysen-Mykorrhiza. *Flora*, **116**, 1923, p. 405.

DENIS, MARCEL (1919). Sur quelques thalles d'*Aneura* dépourvus de chlorophylle. *C. R. Acad. des Sc.* **168**, 1919, p. 64.

DIES (1899). Die endotrophe Mykorrhiza von *Podocarpus* u. ihre physiologische Bedeutung. *Landwirtsch. Versuchsstation*, **51**, 1899, p. 241.

DRUDE, O. (1873). *Die Biologie von* Monotropa Hypopitys *L. u.* Neottia Nidus-avis *L. unter vergleichender Hinzuziehung anderer Orchideen.* Eine von d. Philos, Facult. der Georg-Augustus Universität zu Göttingen gekrönte Preisschrift. Göttingen, 1873.

DUBOIS, R. (1925 *a*). Les plantes vertes sans racines et la culture par symbiose. *C. R. Soc. de Biol.* **93**, 1925, p. 1487.

—— (1925 *b*). Sur le mode d'alimentation des Broméliacées sans racines. *C. R. Acad. d. Sc.* **180**, 1925, p. 1050.

—— (1925 *c*). Les fleurs de l'air. *La Nature*, **53**, 1925, p. 307.

DUCHARTRE, P. (1846). Note sur *L'Hypopitys multiflora*. *Ann. d. Sc. nat. Bot.* 3me sér. **6**, 1846, p. 29.

DUCOMET, V. (1909). Contribution à l'étude de la maladie du Châtaignier. *Ann. École nat. Agric. de Rennes*, **3**, 1909.

Ducomet, V. (1923–24). Théorie mycorhizienne et selection. *Ann. École d'Agric. de Grignon*, 8, 1923–24.

Dufrénoy, J. (1917). The endotrophic mycorrhiza of the Ericaceae. *New Phytologist*, 16, 1917, p. 222.

—— (1920). Actinomyces-like endotrophic mycorrhiza. *New Phytologist*, 19, 1920, p. 40. (See also *Bot. Abs.* 8, 1289, p. 186.)

—— (1922). Sur la tuméfaction et la tubérisation. *C. R. Acad. d. Sc.* 174, 1922, p. 1725. (See also *Review of App. Mycol.* 2, 1923, p. 135.)

—— (1922). Les cellules polynuclées des mycorhizes de *Châtaignier*. *C. R. Soc. de Biol.* 86, 1922, p. 535.

Duggar, B. M. (1905). The principles of mushroom growing and mushroom spawn making. *U.S. Dept. of Agric. Bur. of Pll. Industry, Bull.* 85, 1905.

Duggar, B. M. and Davis, A. R. (1916). Studies in the physiology of the fungi. I. Nitrogen fixation. *Ann. of the Missouri Bot. Garden*, 3, 1916, p. 413.

Duhamel du Monceau (1758). *La physique des arbres.* Paris, 1758.

Dunham, E. M. (1916). Fungus spores in a moss capsule. *The Bryologist*, 19, 1916, p. 89.

Ebermayer, Ernst. (1888). Warum enthalten die Waldbäume keine Nitrate? *Ber. d. d. bot. Gessel.* 6, 1888, p. 217.

Ehrenberg (1820). Epistoles de Mycetogenesi. *Nov. Act. Acad. Nat.* 10, 1820.

Eidam, E. (1879). Ueber Pilzentwicklung in den Wurzeln der Orchideen. *Jahresber. d. bot. Section d. Schles. Gesell. f. vaterländische Kultur*, 57, 1879, p. 297. (See also *Bot. Centralbl.* 8, 1881, p. 2.)

Elenkin, A. A. (1907). Die Symbiose von Gesichtspunkt des beweglichen Gleichgewiches der zusammenlebenden Organismen betrachtet. *Jahrb. f. Pflanzenkrankheit*, 1, 1907.

Fabre, I. H. (1855). Recherches sur les tubercules de l'*Himanthoglossum hircinum*. *Ann. d. Sc. nat. Bot.* 4me sér. 3, 1855.

—— (1856). De la germination des Ophrydées et de la nature de leur tubercules. *Ann. d. Sc. nat. Bot.* 4me sér. 5, 1856.

Falck, R. (1923). Erweiterte Denkschrift über die Bedeutung der Fadenpilze für die Nutzbarmachung der Abfallstoffe zur Baumernährung im Walde u. über de Möglichkeit einer nachträglichen pilzlichen Aufschliessung des Trockentorfs. *Mykol. Unters. u. Ber.* 2. Cassel, 1923.

Fankhauser, L. (1873). Ueber die Vorkeim von *Lycopodium*. *Bot. Zeitung*, 31, 1873, p. 1.

Farmer, J. B. and Freeman, W. G. (1899). On the structure and affinities of *Helminthostachys zeylanica*. *Ann. of Botany*, 13, 1899, p. 421.

Figdor, H. (1896–97). Ueber *Cotylanthera* Bl. Ein Beitrag zu Kenntniss tropischer Saprophyten. *Ann. du Jard. Bot. de Buitenzorg*, 14, 1896–97, p. 213.

Frank, A. B. (1880). *Die Krankheiten der Pflanzen.* Breslau, 1880. p. 648.

—— (1885 *a*). Ueber die Ernährung gewisser Bäume durch Pilze. *Ber. d. d. bot. Gesell.* 3, 1885, p. 128.

—— (1885 *b*). Neue Mittheilungen ueber die Mykorrhiza der Bäume u. der *Monotropa Hypopitys*. *Ber. d. d. bot. Gesell.* 3, 1885, p. xxvii.

—— (1887 *a*). Ueber Ursprung u. Schicksal der Salpetersäure in den Pflanze. *Ber. d. d. bot. Gesell.* 5, 1887, p. 472.

—— (1887 *b*). Ueber neue Mykorrhiza-formen. *Ber. d. d. bot. Gesell.* 5, 1887, p. 395.

—— (1887 *c*). Sind die Wurzelschwellungen der Erlen u. Eleägnaceen, Pilzgallen? *Ber. d. d. bot. Gesell.* 5, 1887, p. 50.

—— (1888 *a*). Ueber die physiologische Bedeutung der Mykorrhiza. *Ber. d. d. bot. Gesell.* 6, 1888, p. 248.

—— (1888 *b*). Ueber den Einfluss welchen das Sterilisiren des Erdbodens auf die Pflanzenentwickelung ausübt. *Ber. d. d. bot. Gesell.* 6, 1888, p. lxxxvii.

—— (1888 *c*). *Untersuchungen ueber die Ernährung des Pflanzen mit Stickstoff.* Berlin, 1888. p. 116.

FRANK, A. B. (1890). Ueber Assimilation von Stickstoff aus der Luft durch *Robinia Pseudoacacia*. *Ber. d. d. bot. Gesell.* **8**, 1890, p. 292.

—— (1891). Ueber die auf Verdauung von Pilzen abzielende Symbiose des mit endotrophen Mykorrhizen begabten Pflanzen sowie der Leguminosen u. Erlen. *Ber. d. d. bot. Gesell.* **9**, 1891, p. 244.

—— (1892 *a*). Die Ernährung der Kiefer durch ihre Mykorrhiza-Pilze. *Ber. d. d. bot. Gesell.* **10**, 1892, p. 577.

—— (1892 *b*). *Lehrbuch der Botanik*. Leipzig, 1892. pp. 255–274.

—— (1892 *c*). Die Pilzsymbiose der Leguminosen. *Landwirthschaft. Jahrb.* **21**. Berlin, 1892.

—— (1894). Die Bedeutung der Mykorrhizapilze für die gemeine Kiefer. *Forstwiss. Centralbl.* **16**, 1894. (See also *Bot. Centralbl.* **62**, 1895, p. 18.)

FRANK, A. B. und OTTO, R. (1890). Untersuchungen ueber Stickstoffassimilation in der Pflanzen. *Ber. d. d. bot. Gesell.* **8**, 1890, p. 331.

FREEMAN, E. M. (1903). The seed fungus of *Lolium temulentum* L. *Phil. Trans. Roy. Soc. London*, **196**, B, 1903.

—— (1904). Symbiosis in the genus *Lolium*. *Minnesota Botanical Studies*, 1904.

—— (1906). The affinities of the fungus of *Lolium temulentum* L. *Annales Mycologici*, **4**, 1906.

FRIES, E. (1874). *Hymenomycetes europaei*. Upsaliae, 1874.

FUCHS, A. und ZIEGENSPECK, H. (1923). *Mycorrhiza* und Boden. *Bot. Arch.* **3**, 1923, p. 237.

—— —— (1924 *a*). Die Pilzverdauung der Orchideen. *Bot. Arch.* **6**, 1924, p. 193.

—— —— (1924 *b*). Entwicklungsgeschichte einiger deutschen Orchideen. *Bot. Arch.* **5**, 1924, p. 120.

—— —— (1924 *c*). *Orchis Traunsteineri* Sauter. Sonderab. aus dem *Berichte d. Naturwissenschaftl. Vereins f. Schwaben u. Neuberg*, **43**, 1924, p. 5.

—— —— (1925). Bau und Form der Wurzeln der einheimischen Orchideen im Hinblick auf ihre Aufgabe. *Bot. Arch.* **12**, 1925, p. 290.

FUCHS, J. (1911 *a*). *Über die Beziehung von Agaricineen und anderen humusbewohnenden Pilzen zur Mykorrhizabildung der Waldbäume*. Separatabdruck aus *Bibl. Bot.* **18**. Stuttgart, 1911.

—— (1911 *b*). Beiträge zur Kenntniss des *Loliumpilzes*. *Hedwigia*, **51**, 1911–12, p. 221.

GALLAUD, I. (1904 *a*). Sur la nature des mycorhizes endotrophiques. *C. R. Soc. Biol.* **56**, 1904, p. 307.

—— (1904 *b*). De la place systématique des endophytes d'Orchidées. *C. R. Acad. d. Sc.* **138**, 1904, p. 513.

—— (1905). Études sur les mycorhizes endotrophes. *Rev. gén. de Bot.* **17**, 1905, pp. 5, 66, 123, 223, 313, 423, 479.

GARJEANNE, A. J. M. (1903). Über die Mykorrhiza der Lebermoose. *Bot. Centralbl.* **15**, 1903, p. 471.

GASPARRINI, G. (1856). *Ricerche sulla natura dei succiatori e la escrezione delle radici*. Naples, 1856. p. 23. (See also Schwartz, *Unters. aus dem bot. Institut in Tübingen*, **1**, p. 176.)

GERLACH und VOGEL (1902). Stickstoffsammelnde Bakterien. *Centralbl. f. Bakt.* Abt. II, **9**, 1902, pp. 669, 817, 881.

GIBELLI, G. (1883). Nuovi studi sulla malattia del Castagno detta dell' inchiostro. *Acad. Bologna*, **4**, 1883.

GOEBEL, K. (1887). Ueber Prothallien u. Keimpflanzen von *Lycopodum inundatum*. *Bot. Zeitung*, Jahrg. **45**, 1887, cols. 161, 177.

—— (1901). Morphol. und biolog. Bemerkungen. 9. Zur Biologie der Malaxiden. *Flora*, **88**, 1901, p. 94.

—— (1905). *Organography of Plants* (Eng. ed.), Part II, 1905, p. 254.

GOLENKIN, M. (1902). Die Mycorrhiza-ähnlichen Bildungen der Marchantiaceen. *Flora*, **90**, 1902, p. 209.

GÖTTSCHE, C. M. (1843). Anatomisch-physiologische Untersuchungen über *Haplomitrium Hookeri*, mit Vergleichung anderer Lebermoose. *Verhandl. der Kaiserl. Leopol.-Car. Akad. der Naturforscher*, Breslau u. Bonn. **12**, 1843, p. 267.

—— (1858). Uebersicht u. kritische Würdigung der Leistungen in der Hepaticologie. *Bot. Zeitung* (Supplement), 1858. Beilage, pp. 1, 40; Schriften, Aufsätze u. Notizen physiologischen Inhalts.

GRAVIS, A. (1879). Observations anatomiques sur les excroissances des racines de l'aune. *Bull. Soc. Roy. de Bot. de Belgique*, **18**, 1879, p. 50. (See also *Ber. d. d. bot. Gesell.* **3**, 1885, p. 177.)

GREVILLIUS, A. V. (1895). Ueber Mykorrhiza bei der Gattung *Botrychium*, etc. *Flora*, **80**, 1895, p. 445.

GROOM, PERCY (1894). Contributions to the knowledge of monocotyledonous saprophytes. *Journal Linnean Soc. London, Bot.* **31**, 1894, p. 149.

GROSGLICK, S. (1886). Die Mycorhiza. *Bot. Centralbl.* **25**, 1886, p. 136.

—— (1895). Thismia Aseroë and its Mycorhiza. *Ann. of Botany*, **9**, 1895, p. 327.

GRUENBERG, B. C. Some aspects of the mycorhiza problem. *Bull. Torrey Bot. Club*, **36**, 1909, p. 165.

GUÉRIN, P. (1898). A propos de la présence d'un champignon dans l'ivraie (*Lolium temulentum*). *Journ. de Botanique*, **12**, 1898, p. 384.

GYÖRFFY, I. (1911). *Cladosporium herbarum* on *Buxbaumia viridis*. *The Bryologist*, **14**, 1911, p. 41.

HAGEN, O. (1910). Untersuchungen über norwegische Mucorineen, II. *Vidensk.-Selsk. Skrift*, 1, Mat.-nat. Kl. **17**, 1910.

HAMMARLUND, C. (1923). *Boletus elegans* und *Larix*-Mykorrhiza. *Bot. Notiser*, 1923, p. 305.

HANAUSEK, T. T. (1898). Vorläufige Mittheilung über den von A. Vogl in der Frucht von *Lolium temulentum* entdeckten Pilz. *Ber. d. d. bot. Gesell.* **16**, 1898, p. 203.

HANSTEEN CRANNER, B. (1922). Zur Biochemie u. Physiologie der Grenzschichten lebender Pflanzenzellen. *Meld. fr. Norges Landbrukshøiskole*, **2**, 1922.

HARTIG, R. (1886). [Vortrag.] Über die symbiotischen Erscheinungen im Pflanzenleben. *Bot. Centralbl.* **25**, 1886, p. 350.

—— (1888 *a*). *Untersuchungen aus d. forstbot. Institut zu München.* Berlin 1888.

—— (1888 *b*). Die pflanzlichen Wurzelparasiten. *Allgem. Forst. u. Jagd-Ztg.* 1888, p. 118. (See also *Centralbl. f. Bakt.* **3**, 1888, pp. 19, 58, 91, 118.)

—— (1889). *Lehrbuch der Baumkrankheiten*, 2 Aufl. 1889.

—— (1891). *Lehrbuch der Anatomie u. Physiologie der Pflanzen.* Berlin, 1891.

HARTIG, TH. (1840–51). *Vollständige Naturgeschichte*, 1840–51.

—— (1863). Ueber das sogenannte Absterben der Haarwurzeln. *Bot. Zeitung*, 1863, p. 289.

—— (1878). *Anatomie u. Physiologie der Holzpflanzen.* Berlin, 1878.

HEINRICHER, E. (1900). Unsere einheimischen *Polygala*-arten sind keine Parasiten. *Ber. d. naturwissensch.-medizin Vereins in Innsbruck*, **26**, 1900–1.

HELLRIEGEL, H. und WILLFARTH, H. (1888). Untersuchungen über die Stickstoffnährung der Gramineen u. Leguminoseen. *Zeit. des Vereins f. Rübenzucker-Industrie*, 1888.

HENDERSON, M. W. (1919). A comparative study of the structure and saprophytism of the Pyrolaceae and Monotropaceae with reference to their derivation from the Ericaceae. *Contrib. Bot. Lab. Univ. Pennsylvania*, 1919, p. 42.

HENSCHEL, G. (1887). Ist die zu Mykorrhizabildungen führende Symbiose jungen Fichtenpflanzen schädlich? *Oesterr. Vierteljahresschrift für Forstwesen*, 1887, p. 113. (See also *Just's Jahresber.* 1887, p. 523.)

HESSELMAN, H. (1900). Om mykorrhizabilningar hos arktiska växter. *Ih. Svenska Vet.-Akad. Handl.* **26**, 1900.

HESSELMAN, H. (1917 *a*). Studier över salpeterbildningen i naturliga jordmåner och dess betydelse i växtekologiskt avseende. *Meddel. fr. Stat. Skogsförs-anst.* **13–14**. Stockholm, 1917.

—— (1917 *b*). Om våra skogsföryngringsåtgärders inverkan på salpeterbildningen i marken och dess betydelse för barrskoyens föryngring. *Meddel. fr. Stat. Skogsförs-anst.* **13–14**. Stockholm, 1917.

—— (1917 *c*). Studier över de norrländska tallhedarnas föryngringsvillkor, II. *Meddel. fr. Stat. Skogsförs-anst.* **13–14**. Stockholm, 1917.

HILTNER, L. (1896). Über die Bedeutung des Wurzelknöllchen von *Alnus*... für die Stickstoff Sammlung dieser Pflanze. *Landwirtsch. Versuchsstation*, **46**, 1896, p. 153.

—— (1899). Über die Assimilation des frein atmosphärischen Stickstoffs durch im oberirdischen Pflanzenteilen lebende Mycelien. *Centralbl. f. Bakt. u. Par.* **5**, 1899.

—— (1900). Ueber die Ursachen, welche die Grosse, Zahl, Stellung u. Wirkung der Wurzelknollchen der Leguminosen bedingen. *Arb. d. Biol. Abt. f. Land- u. Forstwirtsch. a. Kaiserl. Ges.-Amt*, **1**, 1900.

HILTNER, L. und NOBBE, F. (1904). Ueber das Stickstoffssammlungsvermögen der Erlen u. Eleagnaceae. *Naturwiss. Zeitschr. f. Land- u. Forstwirtschaft.* **2**, 1904, p. 366.

HOAGLAND, D. R. (1918). The relation of the plant to the reaction of the nutrient solution. *Science*, **48**, 1918.

HOFMEISTER, W. (1851, 1862). *Vergleichende Untersuchungen höherer Kryptogamen*, 1851. *The Higher Crytogamia*. Roy. Society, 1862.

HOLLE, J. G. (1875). Ueber Bau u. Entwickelung der Vegetationsorgane der Ophioglosseen. *Bot. Zeitung*, **33**, 1875, cols. 241, 265, 281, 297, 313.

HOLLOWAY, J. E. (1909). A Comparative Study of the anatomy of six New Zealand species of *Lycopodium*. *Trans. and Proc. N.Z. Inst.* **42**, 1909, p. 356.

—— (1916). Studies in the N.Z. species of the genus *Lycopodium*. Part I. *Trans. and Proc. N.Z. Inst.* **48**, 1916, p. 253.

—— (1917 *a*). Studies in the N.Z. species of the genus *Lycopodium*. Part II. *Trans. and Proc. N.Z. Inst.* **49**, 1917, p. 80.

—— (1917 *b*). The Prothallus and young plant of *Tmesipteris*. *Trans. and Proc. N.Z. Inst.* **50**, 1917, p. 1.

—— (1920). Studies in the N.Z. species of the genus *Lycopodium*. Part IV. The structure of the prothallus in five species. *Trans. and Proc. N.Z. Inst.* **52**, 1920, p. 193. (See also *Bot. Abs.* **10**, 1826, 1921–22, p. 277.)

HOLM, T. (1897). *Oblaria virginica*. A morphological and anatomical study. *Ann. of Botany*, **11**, 1897, p. 369.

HOOKER, J. (1874). Address to Section of Botany. *Report Brit. Ass. f. Adv. of Sc.* Belfast, 1874.

HOOKER, W. J. (1825). *Monotropa uniflora*. In Hooker's *Exotic Flora*, **2**, 1825, p. 85 (with plate).

HOPPE-SEYLER, F. (1889). Ueber Huminsubstanzen, ihre Enstehung- u. ihre Eigenschaften. *Zeitschr. für physiol. Chemie*, **13**, 1889, p. 66.

HÖVELER, W. (1892). Ueber die Verwerthung des Humus bei der Ernährung der Chlorophyllfuhrenden Pflanzen. *Jahrb. f. wiss Bot.* **24**, 1892, p. 283.

HUBER, B. (1921). Zur Biologie der Torfmoose Orchidee, *Liparis Loeselii*. *Sitzungsber. d. k. Akad. d. Wiss. Wien*, **130**, 1921, p. 307.

HUMPHREY, H. B. (1906). Life history of *Fossombronia longiseta*. *Ann. of Botany*, **20**, 1906, p. 83.

IRMISCH, TH. (1847). Beschreibung des Rhizoms von *Sturmia Looselii*. *Bot. Zeitung*, Jahrg. 5, 1847, col. 137.

—— (1850). *Zur Morphologie der monokotylischen Knollen in Zwiebelgewächse*. Berlin, 1850. p. 159.

—— (1853). *Beiträge zur Biologie und Morphologie der Orchideen*. Leipzig, 1853.

IRMISCH, TH. (1854). Bemerkungen über *Malaxis paludosa*. *Flora*, N.R. Jahrg **12**, 1854, p. 625.

—— (1863). Ein kleiner Beitrag zur Naturgeschicht der *Microstylis monophylla*. *Flora*, **21**, 1863, p. 1.

JACCARD (1904). Les mycorhizes et leur rôle dans le nutrition des essences forestières. *Journal forestier Suisse*, Feb. 1904.

JANCZEWSKI, E. (1874). Das Spitzenwachsthum der Phanerogamenwurzeln. *Bot. Zeitung*, **32**, 1874, col. 113.

—— (1875). Recherches sur le développement des radicelles dans les Phanérogames. *Ann. d. Sc. nat. Bot.* 5me sér. **20**, 1875.

JANSE, J. M. (1896–97). Les endophytes radicaux de quelques plantes Javanaises. *Ann. du Jard. bot. de Buitenzorg*, **14**, 1896–97, p. 53.

—— (1876–97). Quelques mots sur le développement d'une petite Truffle. *Ann. du Jard. bot. de Buitenzorg*, **14**, 1896–97, p. 202.

JEFFERIES, T. A. The vegetative anatomy of *Molinia coerulea*. *New Phytologist*, **15**, 1916, p. 49.

JEFFREY, E. C. (1898). The gametophyte of *Botrychium Virginianum*. *Proc. Canad. Institute*, **5**, 1898.

JENNINGS, A. V. and HANNA, H. (1898). *Corallorhiza innata* and its mycorhiza. *Sc. Proc. of the Roy. Dublin Soc.* **9**, 1898, p. 1.

JOHOW, F. (1885). Die chlorophyllfreien Humusbewohner West-Indiens, biologisch-morphologisch dargestellt. *Prings. Jahrb. f. wiss. Bot.* **16**, 1885, p. 415.

—— (1887). Die chlorophyllfreien Humuspflanzen nach ihren biologischen und anatomisch-entwickelungs-geschlichten Verhaltnissen. *Prings. Jahrb. f. wiss. Bot.* **20**, 1889, p. 475.

JONES, F. R. (1924). A mycorrhizal fungus in the roots of legumes and some other plants. *Journ. Agric. Research*, **29**, 1924, p. 459.

JUMELLE, H. (1905). Influences des endophytes sur la tubérisation des *Solanum*. *Rev. gén. de Bot.* **17**, 1905, p. 49.

KAMIENSKI, F. (1881). Die Vegetationsorgane der *Monotropa Hypopitys* L. *Bot. Zeitung*, **29**, 1881, p. 458.

—— (1882). Les organes vegetatifs du *Monotropa Hypopitys* L. Extrait des *Mém. de la Soc. nat. d. Sc. nat. et math. de Cherbourg*, **24**, 1882, p. 5.

—— (1886). Ueber symbiotische Vereinigung v. Pilzmycelien mit den Wurzeln höherer Pflanzen. *Arb. d. St Petersburg. Naturf.-Gesell.* **17**, 1886, p. 34. (For summary, see *Bot. Centralbl.* 1887, p. 2.)

KAMERLING, Z. (1911). Over het voorkomen van Wortelknolletjes big *Casuarina equisetifolia*. *Natuurkundige Tydschrift v. Ned. Indie*, **71**, 1911, p. 20.

KAUFFMAN, C. H. (1906). *Cortinarius* as a mycorhiza-producing fungus. *Bot. Gazette*, **42**, 1906, p. 209.

KERNER VON MARILAUN (1887). *Pflanzenleben*. Leipzig, 1887–91. **1**, p. 92.

KILLIAN, CH. (1923–24). Cultures d'Hépatiques. *C. R. Soc. de Biol.* **88**, 1923; **91**, 1924.

KNIEP, H. (1915). Beiträge zur Kenntniss der Hymenomyzeten. III. Über die konjugierten Teilungen und die phylogenitische Bedeutung der Schnallenbildungen. *Zeitschr. f. Bot.* **7**, 1915.

KNUDSON, L. (1922). Non-symbiotic germination of Orchid seeds. *Bot. Gazette*, **73**, 1922, p. 1.

—— (1924). Further Observations on non-symbiotic germination of Orchid seeds. *Bot. Gazette*, **77**, 1924, p. 212.

—— (1925). Physiological Study of the symbiotic germination of Orchid seeds. *Bot. Gazette*, **79**, 1925, p. 345.

KNY, L. (1879). Ueber eigenthümliche Durchwachsungen an den Wurzelhaaren zweier Marchantiaceen, untersucht von Dr Böttger. *Verhandl. (u. Sitzungsber.) d. bot. Vereins d. Prov. Brandenburg*, **21**, 1879, p. 2. (See also *Bot. Zeitung*, **37**, 1879, p. 450.)

KOCH, L. (1882). Die Entwicklung des Samens von *Monotropa Hypopithys* L. *Prings. Jahrb. f. wiss. Bot.* 13, 1882, p. 202.

—— (1887). Ueber die directe Ausnutzung vegetabilischer Reste durch chlorophyllhaltigen Pflanzen. *Ber. d. d. bot. Gesell.* 5, 1887, p. 350.

KÖPPEN, F. T. (1889). *Geographische Verbreitung der Holzgewächse des europäischen Russlands und des Kaukasus.* 2. Beiträge z. Kenntnis des Russ. Reiches. St Petersburg, 1889.

KRAMAR, N. (1899). Studie über die Mykorrhiza von *Pirola rotundifolia. Bull. internat. de l'Acad. des Sc. de Bohême*, 1899.

KÜHN, V. R. (1889). Untersuchungen über die Anatomie der Marattiaceen. *Flora*, 47, 1889, p. 457.

KUSANO, S. (1911). *Gastrodia elata* and *Armillaria mellea. Journ. of College of Agric. Univ. of Tokio*, 4, 1911. (See also *Ann. of Botany*, 25, 1911, p. 522.)

LAING, E. V. (1923). Tree roots: their action and development. *Trans. of Roy. Scottish Arbor. Soc.* 37, 1923, p. 6.

LANG, W. (1899). The prothallus of *Lycopodium clavatum* L. *Ann. of Botany*, 13, 1899, p. 279.

—— (1901). Preliminary statement on the prothalli of *Ophioglossum pendulum, Helminthostachys zeylanica* and *Psilotum* sp. *Proc. Roy. Soc. London*, 68, B. 1901, p. 405.

—— (1902). On the prothalli of *Ophioglossum pendulum* and *Helminthostachys zeylanica. Ann. of Botany*, 16, 1902, p. 23.

—— (1904). On a prothallus provisionally referred to *Psilotum. Ann. of Botany*, 18, 1904, p. 571.

LANGE, J. E. (1923). Studies in the Agarics of Denmark. *Dansk Bot. Ark.* 4, 1923.

LAURENT, E. (1888). Recherches experimentales sur la formation d'amidon dans les plantes aux dépens de solutions organiques. *Bull. Soc. Roy. de Bot. de Belgique*, 26, 1888.

—— (1891). Recherches sur les nodosités radicales des Légumineuses. *Ann. de l'Institut Pasteur*, 5, 1891, p. 105.

LAWSON, A. A. (1917). The prothallus of *Tmesipteris Tannensis. Trans. Roy Soc. Edinburgh*, 51, 1917, p. 785.

—— (1918). The gametophyte generation of the Psilotaceae. *Trans. Roy. Soc. Edinburgh*, 52, 1918, p. 93.

LECOMTE, H. (1887). Note sur le mycorhize. *Bull. de la Soc. bot. de France*, 34, 1887, p. 38.

LEES, EDWIN (1844). Parasitic growth of *Monotropa Hypopitys. The Phytologist*, 1, 1844, p. 97.

LEGROUX et MAGROU, J. (1920). *Ann. de l'Institut Pasteur*, 34, 1920.

LEITGEB, H. (1865). Die Luftwurzeln der Orchideen. *Denkschrft. d. Akad. wiss. Wien*, 1865.

—— (1874–81). *Untersuchungen über die Lebermoose*, 6 vols. 1874–81.

LENDNER, A. (1896). Contribution à la connaissance des mycorhizes des Orchidées par Wahrlich. Abstract, *Rev. Mycolog.* 20, 1896, p. 1.

LEPESCHIN, W. (1924). The influence of vitamins upon the development of yeasts and molds. Contribution to the bios problem. *Amer. Journ. of Bot.* 11, 1924, p. 164.

LEWIS, F. J. (1924). An endotropic fungus in the Coniferae. *Nature*, 114, 1924, p. 860.

LIEFE, D. (1924). Der heutige Stand der Mycorrhizaforschung. *Berk. f. Forst- u. Jagdwesen*, 56, 1924, p. 747.

LIFE, A. C. (1901). The tubercle-like roots of *Cycas revoluta. Bot. Gazette*, 31, 1901, p. 265.

LINDSAY, J. (1794). Account of the germination and raising of ferns from the seed. *Trans. Linn. Soc. London*, 2, 1794, p. 93.

LINK, H. F. (1840). Keimung von *Goodyera procera* und *Angraecum maculatum. Icones selectae anatomico-botanicae.* fasc. II, Tab. VII, 1840, p. 10.

Löhnis, F. and Hansen (1921). Nodule bacteria of leguminous plants. *Journ. Agric. Research*, **20**, 1921, p. 543.

Lundström, A. M. (1889). Einige Beobachtungen über *Calypso borealis*. *Bot. Centralbl.* **38**, 1889, p. 697

Luxford, G. (1844). Botanical notes. *Monotropa Hypopitys*. *The Phytologist*, **1**, 1844, p. 43.

Lyon, H. L. (1905). A new genus of Ophioglossaceae. *Bot. Gazette*, **40**, 1905, p. 455.

MacDougal, D. T. (1898 *a*). The mycorhiza of *Aplectrum*. *Bull. Torrey Bot. Club*, **25**, 1898, p. 110.

—— (1898 *b*). Saprophytism. *Plant World*, **2**, 1898, p. 23.

—— (1899 *a*). Significance of mycorhiza. *Biol. Lect. from Marine Biol. Lab., Wood's Hole*, 1899, p. 49.

—— (1899 *b*). Symbiotic Saprophytism. *Ann. of Botany*, **13**, 1899, p. 1

—— (1899 *c*). Symbiosis and Saprophytism. *Contrib. from New York Bot. Garden*, No. 1, 1899. (Reprinted from *Bull. Torrey Bot. Club*, **26**, 1899, p. 511.)

Magnus, P. (1877). Durch Parasiten veranlasste Wurzelauswüchse. *Verhandl. (u. Sitzungsber.) d. bot. Vereins Prov. Brandenburg*, 1877, p. 79.

Magnus, W. (1900). Studien an der endotrophen Mykorrhiza von *Neottia Nidus-avis*. *Jahr. f. wiss. Bot.* **35**, 1900, p. 205.

—— (1911). *Mycorrhiza*. Botan. Wandtafeln mit erläuterndem Text von L. Kny. Abt. 13, 1911, Tafeln 116 u. 117, p. 525.

Magrou, J. (1921). Symbiose et tubérisation. *Ann. d. Sc. nat. Bot.* 9me sér **3**, 1921, p. 181. (See also *Bot. Abs.* **11**, 3083, 1922, p. 3.)

—— (1922). La symbiose chez les plantes. *Bull. Inst. Pasteur*, **20**, 1922, p. 169.

—— (1924 *a*). L'immunité humorale chez les plantes. *Rev. path. vég. et Ent. agric.* **11**, 1924, p. 189.

—— (1924 *b*). A propos du pouvoir fungicide des tubercules d'Ophrydées. *Ann. d. Sc. nat. Bot.* 10me sér. 1924, p. 265.

—— (1924 *c*). Remarques sur les cultures expérimentales de Pommes de terre avec endophyte. *Ann. d. Sc. nat. Bot.* 10me sér. **6**, 1924, p. 265.

—— (1925). La symbiose chez les Hépatiques. Le *Pellia epiphylla* et son champignon commensal. *Ann. d. Sc. nat. Bot.* 10me sér. **7**, 1925, p. 725.

Malpighi, M. (1675, 1679). *Anatome plantarum*. London, Part I, 1675; Part II, 1679.

—— (1686). *Opera omnia*. Part II, 1686, pp. 69–71, Figs. 2, 4.

Mangin, L. (1898). Sur la structure des mycorhizes. *C. R. Acad. des Sc.* **126**, 1898, p. 978.

—— (1910). Introduction à l'étude des mycorhizes des arbres forestiers. *Nouvelles Archives du Muséum d'Hist. Nat. de Paris*, 5me sér. **2**, 1910, p. 245.

Marcuse, M. (1902). *Anatom-biolog. Beiträge zur Mykorrhizenfrage*. Inaug. Dissertation, Universität Jena. 1902.

Masui, Koki (1926 *a*). A study of the mycorrhiza of *Abies firma* S. and Z., with special reference to its mycorrhizal fungus, *Cantharellus floccosus* Schw. *Mem. of the Coll. of Sc. Kyoto Imp. Univ.* **2**, Ser. B, 1926, pp. 15, 85.

—— (1926). A study of the ectotrophic mycorrhiza of *Alnus*. *Mem. of the Coll. of Sc. Kyoto Imp. Univ.* **2**, Ser. B, 1926, p. 190.

—— (1926). The compound mycorrhiza of *Quercus pausidenta* Fr. *Mem. of the Coll. of Sc. Kyoto Imp. Univ.* **2**, Ser. B, 1926, p. 162.

Mattirolo, O. (1887). Sul parassitismo dei Tartufi e sulla quistione delle Mycorhizae. *Malpighia*, fasc. 8–9. Turin, 1887.

McDougall, W. B. (1914). On the mycorhiza of forest trees. *Amer. Journ. of Bot.* **1**, 1914, p. 51.

—— (1916). The growth of forest tree roots. *Amer. Journ. of Bot.* **3**, 1916, p. 384.

McDougall, W. B. (1918). Classification of symbiotic phenomena. *The Plant World*, **21**, 1918, p. 250.

—— (1922 *a*). Symbiosis in a deciduous forest. I. *Bot. Gazette*, **73**, 1922, p. 200.

—— (1922 *b*). Mycorhizas of forest trees. *Journ. of Forestry*, **20**, 1922, p. 255.

McLennan, E. (1920). The endophytic fungus of *Lolium*. Part I. *Proc. Roy. Soc. Victoria*, **32** (N.S.), Part II, 1920, p. 251.

—— (1926). The endophytic fungus of *Lolium*. II. The mycorrhiza in the roots of *L. temulentum*. *Ann. of Botany*, **40**, 1926, p. 43.

McLuckie, J. (1922 *a*). Studies in symbiosis. I. The mycorhiza of *Dipodium punctatum*. *Proc. Linn. Soc. N.S. Wales*, **47**, 1922, p. 293.

—— (1922 *b*). Studies in symbiosis. II. The apogeotropic roots of *Macrozamia spiralis* and their physiological significance. *Proc. Linn. Soc. N.S. Wales*, **37**, 1922, p. 319.

—— (1923 *a*). Studies in symbiosis. III. Contributions to the morphology and physiology of the root nodules of *Podocarpus spinulosa* and *P. elata*. *Proc. Linn. Soc. N.S. Wales*, **48**, 1923, p. 82.

—— (1923 *b*). Studies in symbiosis. IV. The root nodules of *Casuarina cunninghamia* and their physiological significance. *Proc. Linn. Soc. N.S. Wales*, **48**, 1923, p. 194.

—— (1924). Studies in symbiosis. V. A contribution to the physiology of *Gastrodia sesamoides*. *Proc. Linn. Soc. N.S. Wales*, **48**, 1924, p. 436.

Melin, Elias (1917). Studier över de norrländska myrmarkernas vegetation med särkild hausyn till deras skogs-vegetation efter torrlägning. *Akad. Avhandl. Uppsala*, 1917.

—— (1921). Über die Mykorrhizenpilze von *Pinus silvestris* L. and *Picea Abies* L. *Svensk. Bot. Tidskr.* **15**, 1921, p. 192.

—— (1922 *a*). Mykorrhizas of *Pinus silvestris* and *Picea Abies*. *Journ. of Ecology*, **9**, 1922, p. 254.

—— (1922 *b*). Untersuchungen über die *Larix*-Mykorrhiza. I. Synthese der Mykorrhiza in Reinkultur. *Svensk. Bot. Tidskr.* **16**, 1922, p. 161.

—— (1922 *c*). *Boletus* Arten als Mykorrhizen-pilze der Waldbäume. *Ber. d. d. bot. Gesell.* **40**, 1922, p. 94.

—— (1923 *a*). Experimentelle Untersuchungen über die Konstitution und Ekologie der Mykorrhizen von *Pinus silvestris* u. *Picea Abies*. *Mykologische Untersuchungen u. Berichte* (Falck), **2**, 1923, p. 73.

—— (1923 *b*). Experimentelle Untersuchungen über die *Birken*- u. *Espen*-mykorrhizen u. ihre Pilzsymbionten. *Svensk. Bot. Tidskr.* **17**, 1923, p. 479.

—— (1924 *a*). Zur Kenntniss der Mykorrhizapilze von *Pinus montana*. *Bot. Notiser*, 1924, p. 69.

—— (1924 *b*). Über den Einfluss der Wasserstoffionen Konzentration auf die Virulenz der Wurzelpilze von Kiefer u. Fichte. *Bot. Notiser*, 1924, p. 38.

—— (1925 *a*). Untersuchungen über die *Larix*-Mykorrhiza. II. Zur weiteren Kenntniss der Pilzsymbionten. *Svensk Bot. Tidskr.* **19**, 1925, p. 98.

—— (1925 *b*). Über die Activität von proteolytischen u. verwandten Enzymen einigen als Mykorrhizenpilze bekannten Hymenomyceten. *Biochem. Zeitschrift*, **157**, 1925, p. 146.

—— (1925 *c*). *Untersuchungen über die Bedeutung der Baum-mykorrhiza*. Jena, 1925.

Mettenius, G. (1856). *Filices Horti. Botanici Lipsiensis*. Leipzig, 1856, p. 119.

Meyen, J. (1829). Ueber das herauswachsen parasitischer Gewächse aus den Wurzeln anderer Pflanzen. *Flora*, **1**, 1829, p. 49.

Meyer, A. (1886). Ueber die Knollen der einheimischen Orchideen. *Arch. d. Pharmacie*, **24**, 1886, p. 3.

Michelettti, L. (1901). Sulla tossicita dei semi di *Lolium temulentum*. *Bull. d. Soc. Ital.* **8**, 1901.

Miehe, Hugo (1918). Anatomical investigations on *Casuarina equisetifolia*, with remarks on mycorhiza problems. *Flora*, **11–12**, 1918, p. 431. (See also *Bot. Abs.* **4**, 248, p. 35.)

MIMURA, S. (1915). Researches on the culture of 'Matsuduke.' Extract from *Bull. of the Forestry Exp. Station*, Tokio, 1915.

MOELLER, H. (1885). Ueber *Plasmodiophora alni*. *Ber. d. d. bot. Gesell.* **3**, 1885, p. 102.

—— (1890). Beitrag zur Kenntniss der *Frankia subtilis*. *Ber. d. d. bot. Gesell.* **8**, 1890, p. 215.

MÖLLER, A. (1902). Ueber die Wurzelbildung der ein- u. zweijahrigen Kiefer im märkischen Sandboden. *Zeitschrift f. Forst- u. Jagdwesen*, **34**, 1902.

—— (1903). Untersuchungen über ein- und zweijahrige Kiefer in märkischen Sandboden. *Zeitschrift f. Forst- und Jagdwesen*, **35**, 1903, p. 257.

—— (1904). Karenzerscheinungen bei der Kiefer. Ein Beitrag zur wissenschaftlichen Begründung einer forstlichen Düngerlehre. *Zeitschrift f. Forst- und Jagdwesen*, **36**, 1904.

—— (1906). Mykorrhizen und Stickstoffernährung. *Ber. d. d. bot. Gesell.* **24**, 1906, p. 230.

MOLLBERG, A. (1884). Untersuchungen über die Pilz in den Wurzeln der Orchideen. *Jenaische Zeitschrift f. Naturwissenschaft*, **17**, 1884, p. 519.

MOLLIARD, M. (1907). Action morphogénique de quelques substances organiques sur les végétaux supérieurs. *Rev. gén. d. Bot.* **19**, 1907, pp. 241, 329, 357.

—— (1915). Production expérimentale de tubercles aux depens de la tige principal chez la Pomme de terre. *C. R. Acad. des. Sc.* **161**, 1915, p. 531.

—— (1920). Tubérisation aseptique de la Carotte et du Dahlia. *C. R. Soc. de Biol.* **83**, 1920.

MOORE and BETCHE (1893). *Flora of New South Wales*. 1893.

MÜLLER, P. E. (1886). Bemerkungen über die Mycorhiza der Buche. *Bot. Centralbl.* **26**, 1886, p. 22.

—— (1887 *a*). *Studien über die naturlichen Humusformen u. deren Einwirkung auf Vegetation u. Boden*. Berlin, 1887. p. 173.

—— (1887 *b*). Om Bjergfyrren, *Pinus montana*. *Tidskrift f. Skovbrug*, **8**, **9**, **11**, 1887.

—— (1889). Recherches sur les formes naturelles de l'humus. *Ann. Soc. agron. français et etrangère*, 1889.

—— (1902). Sur deux formes de mycorhizes chez le Pin de montagne (*P. montana*). *Bull. de l'Acad. roy. des. Sc. et des lettres de Danemark*, **6**, 1902.

—— (1903 *a*). Über das Verhältniss der Bergkiefer zur Fichte in den jütländischen Heidekulturen. *Naturwiss. Zeitschrift f. Land- u. Fortwirtschaft*, 1903, pp. 220, 289, 377.

—— (1903 *b*). Om bjergfyrrens forhold til rødgranen i de jydske Hedeculturer. *Tidskrift f. Skovbrug*, Supplement, Copenhagen, 1903.

MULLER, C. (1848). Recherches sur le développement de l'embryon végétal. *Ann. d. Sc. nat. Bot.* 3me sér. **9**, 1848, p. 33.

NÄGELI, K. (1842). Bot. Beiträge: Pilz im Innern von Zellen. *Linnaea*, **16**, 1842, p. 278. (See also Sorauer, *Pflanzenkrankheiten*, **1**, p. 748.)

NAWASCHIN, S. (1892). Ueber die Brandkrankheit der Torfmoose. *Bull. de l'Acad. des Sc. de St Pétersbourg*, **35**, 1892, p. 531.

NEGER, F. W. (1903). Ein Beitrag zur Mykorrhizafrage. *Centralbl. f. Bakt. u. Par.* **11**, 1903, 350.

NĚMEC, B. (1899). Die Mykorrhiza einiger Lebermoose. *Ber. d. d. bot. Gesell.* **17**, 1899, p. 311.

—— (1901). Ueber die Mykorrhiza bei *Calypogeia trichomanes*. *Beit. z. Bot. Centralbl.* **16**, 1901.

NESTLER, A. (1898). Ueber einen in der Frucht von *Lolium temulentum* vorkommenden Pilz. *Ber. d. d. bot. Gesell.* **16**, 1898, p. 203.

—— (1904). Zur Kenntniss der Symbiose eines Pilzes mit dem Taumelloch. *Sitzungsber. d. k. Akad. d. Wiss.* Wien, **113**, 1904, p. 529.

NEWMAN, E. (1844). Notes on supposed parasitism of *Monotropa Hypopitys*. *The Phytologist*, **1**, 1844, p. 297.

NICOLAS, G. (1924). Formations mycorhiziques dans une Hépatique à thalle (*Lunularia vulgaris*). *C. R. Acad. des Sc.* **178**, 1924, p. 228.

NIKITINSKY, J. (1902). Ueber die Zersetzung der Huminsäure durch physikalisch-chemisch Agention und durch Mikroorganismen. *Prings. Jahrb. f. wiss. Bot.* **37**, 1902, p. 365.

NOACK, R. (1889). Ueber mykorhizenbildende Pilze. Vorläufige Mittheilung. *Bot. Zeitung*, **47**, 1889, p. 389.

NOBBE, F. und HILTNER, L. (1899). Die endotrophe Mykorrhiza von *Podocarpus* und ihre physiologische Bedeutung. *Landwirthsch. Versuchsstation*. Berlin, **51**, 1899, p. 241; **52**, p. 455.

NOBÉCOURT, P. (1923). Sur la production d'anticorps par les tubercles des Ophrydées. *C. R. Acad. des Sc.* **177**, 1923, p. 1055.

NOELLE, W. (1910). Studien zur vergleichenden Anatomie u. Morphologie der Koniferenwurzeln mit Rücksicht auf die Systematik. *Bot. Zeitung*, **68**, 1910, p. 169.

NUTTALL, G. H. (1923). Symbiosis in animals and plants. *Report Brit. Ass. f. Adv. of Sc.*, Liverpool, **91**, 1923, p. 197.

OLIVER, F. W. (1890). On *Sarcodes sanguinea* Torr. *Ann. of Botany*, **4**, 1890, p. 303.

OSBORN, T. G. B. (1909). Lateral roots of *Amyelon radicans* and their mycorrhiza. *Ann. of Botany*, **23**, 1909, p. 603.

PAULSON, R. (1923). The fungus-root (Mycorrhiza). *Essex Nat.* **20**, 1923, p. 177.

—— (1924). Tree mycorrhiza. *Trans. Brit. Mycol. Soc.* **9**, 1924, p. 213.

PEKLO, J. (1903). Einiges über die Mykorrhiza bei den Muscineen. *Bull. intern. Acad. d. Sc. de Bohème*, **8**, 1903.

—— (1906). Zur Lebensgeschichte von *Neottia Nidus-avis*. *Flora*, **96**, 1906, p. 260.

—— (1908). Die epiphytischen Mykorrhizen nach neureren Untersuchungen. I. *Monotropa Hypopitys* L. *Bull. intern. Acad. d. Sc. de Bohème*, 1908.

—— (1909). Beiträge zur Mykorrhiza-problems. *Ber. d. d. bot. Gesell.* **27**, 1909, p. 239.

—— (1910). Die pflanzliche Aktinomykosen. *Centralbl. f. Bakt.* **27**, 1910, p. 451.

—— (1913 *a*). Neue Beiträge zur Lösung des Mykorrhiza problems. *Zeitschrift f. Gär*[illegible] **2**, 1913, p. 246.

—— (1913 [illegible] Zusammensetzung der sogenannten Aleuronschicht. *Ber. d. d. bot. Gesell.* **31**, 1913, p. 370.

PENNINGTON, L. H. (1908). Mycorhiza-producing Basidiomycetes. *Report Mich. Acad. Sci.* **10**, 1908, p. 47.

PENZIG, O. (1885). Die Krankheit der Edelkastanien und B. Frank's Mykorrhiza. *Ber. d. d. bot. Gesell.* **3**, 1885, p. 301.

PESTANA (1907). La maladie des Châtaigniers. *Bull. Soc. Portugaise Sc. nat.* **1**, 1907.

PETRI, L. (1903). Ricerche sul significata morfologico e fisiologico dei prosporoidi (sporangioli di Janse) nelle micorize endotrofiche. *Nuovo Giorn. Bot. Ital.* n. ser., **10**, 1903, p. 541.

—— (1907 *a*). Su le micorize endotrofiche della vite. *Rendic. Accad. Lincei*, 5ª ser., **16**, 1907, p. 789.

—— (1907 *b*). Studi sul marciume della radici nelle viti fillosserate. *Mem. della R. Staz. di Pat. veg. Rome*, 1907.

—— (1908). Rapporto fra micotrofia e attività funzionale nell' olivo. *Rendic. Accad. Lincei*, 5ª ser., **17**, 1908, p. 754.

—— (1915). État actuel des connaissances sur la signification physiologique des mycorhizes des arbres. *Bull. mens. des renseign. agric. et des maladies des plantes*, **6**, 1915, p. 1230.

—— (1918–19). Sopra una presunta malattia parassitaria del proppo. *Annali del R. Ist. sup. forest. naz. di Firenze*, **4**, 1918–19, p. 97.

PEYRONEL, B. (1917). Prime osservazione sulla distribuzione degli Imenomiceti umicoli e sui loro probabali rapporti colle micorize ectotrofiche delle fanerogame. *Rendic. R. Accad. dei Lincei*, 5ª ser., **26**, 1917, p. 326.

—— (1920). Alcuni casi di rapporti Micorizici tra Boletinee ed essenze arboree. *Mem. della R. Staz. di Patologio veg. Rome*, **53**, 1920, p. 24. (See also *Bot. Ab.* **7**, 2169, p. 297.)

—— (1921). Nouveaux cas de rapports mycorhiziques entre Phanérogames et Basidiomycètes. *Bull. de la Soc. Mycol. de France*, **37**, 1921, p. 143.

—— (1922 *a*). Ricerche e studi compiati o in corso presso la R. Staz. di patalogia vegetale. *Bull. mensile R. Staz. Pat. veg.* **111**, 1922, p. 120.

—— (1922 *b*). Nuovi casi di rapporti micorizici tra Basidiomiceti e Fanerogame arboree. *Bull. della Soc. Bot. Ital.* Genoa, **1**, 1922, p. 3.

—— (1922 *c*). Sulla normale presenza di micorize nel grano e in altre piante coltivate e spontanee. *Bull. mens. di inform. e not. della R. Staz. di. Pat. veg. Rome*, **3**, 1922, p. 43.

—— (1923). Fructification de l'endophyte à arbuscules et à vésicules des mycorhizes endotrophes. *Bull. Trim. de la Soc. Mycol. de France*, **39**, 1923, p. 119.

—— (1924). Prime ricerche sulla micorize endotrofiche e sulla microflora radicocola della fanerogame. *Rivista di Biologia*, **5**, **6**, 1924, p. 3.

PFEFFER, W. (1877). Ueber fleischfressende Pflanzen u. ueber die Ernährung durch aufnahme organischer Stoffe ueberhaupt. *Thiel's Landwirtschaftlich Jahrb.* **6**, 1877, p. 997.

—— (1897). *Pflanzenphysiologie*, Eng. ed. 1897. Vol. **1**, p. 359 and Fig. 54.

PFEIFFER, N. E. (1914). Morphology of *Thismia americana*. *Bot. Gazette*, **57**, 1914, p. 122.

PHILLIPS, J. F. V. (1924). *Ocotea bullata*, the "Stinkwood." *S. Af. Journ. of Sc.* **21**, 1924, p. 289.

PIROTTA, R. ed ALBINI, A. (1900). Osservazioni sulla biologia del tartufo giallo (*Terfezia Leonis*). *Rendic. R. Accad. dei Lincei*, **9**, 1900, p. 4.

POULSEN, V. A. (1890). *Triuris major* et Bidrag til Triuridaceernes Naturhistorie. *Bot. Tidsskrift*, **17**, 1890, p. 293.

PRILLIEUX, E. (1856 *a*). De la structure anatomique et du mode de végétation du *Neottia Nidus-avis*. *Ann. d. Sc. nat. Bot.* 4me sér. **5**, 1856, p. 267.

—— (1856 *b*). Observation sur la germination et le développement d'une Orchidée (*Anagraecum maculatum*). *Ann. d. Sc. nat. Bot.* 4me sér. **5**, 1856.

—— (1860). Observation sur la germination de *Miltonia spectabilis* et de divers autres Orchidées. *Ann. d. Sc. nat. Bot.* 4me sér. **13**, 1860.

PRILLIEUX, E. et RIVIÈRE, A. (1856). Observation sur la germination et le développement d'une Orchidée (*Anagraecum maculatum*). *Ann. d. Sc. nat. Bot.* 4me sér. **5**, 1856, p. 267.

RAMANN, E. (1888). Die von Post'schen Arbeiten über Schlamm, Moor, Torf, u. Humus. *Thiel's Landwirtschaftlich Jahrb.* **17**, 1888, pp. 412, 416, 417, 419.

RAMSBOTTOM, J. (1921). Orchid mycorrhiza. Introduction to Messrs Charlesworth and Co.'s *Catalogue*, 1921. (See *Orchid Review*, **30**, 1922, p. 78.)

—— (1922). Orchid mycorrhiza. *Brit. Mycol. Soc. Trans.* **8**, 1922, p. 28.

—— (1922). The germination of Orchid seed, 1922. *Orchid Review*, **30**, 1922, p. 197.

RAYNER, M. C. (1911). Preliminary observations on the ecology of *Calluna vulgaris* on the Wiltshire and Berkshire downs. *New Phytologist*, **10**, 1911, p. 227.

—— (1913). The ecology of *Calluna vulgaris*. *New Phytologist*, **12**, 1913, p. 59.

—— (1915). Obligate symbiosis in *Calluna vulgaris*. *Ann. of Botany*, **29**, 1915, p. 97.

—— (1916). Recent developments in the study of endotrophic mycorhiza. *New Phytologist*, **15**, 1916, p. 161.

—— (1921). The ecology of *Calluna vulgaris*. II. The calcifuge habit. *Journ. of Ecology*, **9**, 1921, p. 60.

RAYNER, M. C. (1922 *a*). Nitrogen fixation in the Ericaceae. *Bot. Gazette*, **73**, 1922, p. 226.

—— (1922 *b*). Mycorrhiza in the Ericaceae. *Trans. of Brit. Mycol. Soc.* **8**, 1922, p. 61.

—— (1923). *Report of the Brit. Ass. f. Adv. of Sc.*, Liverpool, 1923, p. 486.

—— (1925). The nutrition of mycorrhiza plants: *Calluna vulgaris*. *The Brit. Journ. of Exp. Biology*, **2**, 1925, p. 265.

REESS, M. (1871). Ueber die Entstehung der Flechte *Collema glaucescens*. *Monatsber. d. k. preuss. Akad. d. Wiss.* Berlin, 1871, p. 523.

—— (1879). *Ueber die Natur d. Flechten*. Samml. wiss. Vorträge von Virchow. Hft. 320, 1879.

—— (1880). Ueber den Parasitismus von *Elaphomyces granulatus*. *Sitzungsber. d. physik.-med. Soc. z. Erlangen*, 1880; *Bot. Zeitung*, **38**, 1880, p. 729.

—— (1885 *a*). Ueber *Elaphomyces* u. sonstige Wurzelpilze. *Ber. d. d. bot. Gesell.* **3**, 1885, p. 293.

—— (1885 *b*). Weitere Mittheilungen ueber *Elaphomyces granulatus*. *Ber. d. d. bot. Gesell.* **3**, 1885, p. lxiii.

REESS, M. und FISCH, C. (1887). Untersuchungen ueber Bau u. Lebengeschichte der Hirschtrüffel, *Elaphomyces*. *Bibliotheka botanicae*, **7**, 1887, p. 1.

REINITZER, F. I. (1900). Ueber die Eignung der Huminsubstanzen zur Ernährung von Pilzen. *Bot. Zeitung*, **58**, 1900.

REINKE, J. (1873). Zur Kenntniss des Rhizoms von *Corallorhiza* u. *Epipogon*. *Flora*, **31**, 1873, pp. 145, 161, 177, 209.

REINSCH, H. (1852). Ueber einen eigenthümlichen Stoff in der *Monotropa Hypopitys*. *Jahrb. f. praktische Pharmacie u. verwandte Fächer*, **25**, 1852, p. 193.

REISSEK, SIEGFRIED (1847). *Endophyten der Pflanzenzelle*. Naturwissenschaftliche Abhandlungen v. W. Haidinger, 1. Wien, 1847. p. 31.

REXHAUSEN, L. (1920). Über die Bedeutung der ektotrophen Mykorrhiza für die höheren Pflanzen. *Beiträge z. Biologie der Pflanzen*, **14**, 1920, p. 19. (See also *Bot. Abs.* **12**, 3558, 1923, p. 607.)

RIDLER, W. F. F. (1922). The fungus present in *Pellia epiphylla* (L.) Corda. *Ann. of Botany*, **36**, 1922, p. 193.

—— (1923 *a*). Further Observations on the fungus present in *Pellia epiphylla* (L.) Corda. *Ann. of Botany*, **37**, 1923, p. 483.

—— (1923 *b*). The fungus present in *Lunularia cruciata*. *Trans. Brit. Mycol. Soc.* **9**, 1923, p. 82.

RIVETT, M. (1924). The root tubercles in *Arbutus Unedo*. *Ann. of Botany*, **38**, 1924, p. 661.

ROMELL, L. G. (1921). Parallelvorkommen gewisser Boleten und Nadelbäume. *Svensk. Bot. Tidsk.* **15**, 1921, p. 204.

RUSSELL, E. J. (1921). *Soil Conditions and plant growth*. London: Longmans, 1921.

RUSSOW, J. (1872). Vergleichende Untersuchungen der Leitbündel-Kryptogamen. *Mém. Acad. Imp. des Sc. de St Pétersbourg*, **19**, 1872, p. 107.

RYLANDS, T. G. (1844). On the nature of the byssoid substance found investing the roots of *Monotropa Hypopitys*. *The Phytologist*, **1**, 1844, pp. 329, 341.

SADEBECK, R. (1875). *Cohn's Beiträge*, 1875, p. 167.

SALAMAN, R. N. (1924). The inheritance of "cropping" in the potato. *Report of the Imp. Bot. Conference, London*. Cambridge Univ. Press, 1924. p. 40.

SALISBURY, R. A. (1804). On the germination of seeds of Orchidaceae. *Trans. Linn. Soc. London*, **7**, 1804, p. 29 (with plate).

SARAUW, G. F. L. (1892). Rødsymbiose og Mykorrhizer. *Bot. Tidskrift*, **18**, 1892–93, p. 127.

—— (1893). Ueber die Mykorrhizen unserer Waldbäume. *Bot. Centralbl.* **53**, 1893, p. 343.

—— (1903–4). Sur les mycorhizes des arbres forestiers et sur le sens de la symbiose des racines. *Rev. Mycol.* 1903–4.

SCHACHT, H. (1852). *Physiologiche Botanik.* Die Pflanzenzelle. Berlin, 1852.
—— (1853, 1854). Beitrag zur Entwickelungsgeschichte der Wurzel. *Flora*, 17, 1853. (Abstract in *Beitrag z. Anat. u. Phys. der Gewächse*, Berlin, 1854.)
—— (1854 *a*). Ueber Pilzfäden im Innern der Zellen u. der Stärkemehlkörner. *Monatsber. d. Berl. Akad. d. Wiss.* 1854, p. 377. (Abstract in *Flora*, 1854, p. 618.)
—— (1854 *b*). Beiträge zur Anatomie u. Physiologie der Gewächse. IV. Zur Entwicklungsgeschichte der *Monotropa*. VII. Ueber die Fortpflanzung der deutschen Orchideen. *Monatsber. d. Berl. Akad. d. Wiss.* 1854.
—— (1860). *Der Baum.* 2 Aufl. Berlin, 1860.
SCHATZ, W. (1910). *Beiträge zur Biologie der Mykorrhizen.* Inaug. Dissert. Jena, 1910.
SCHNELLENBERG, H. C. (1908). (See Schröter, 1908, p. 305.)
SCHIMPER, PH. W. (1858). *Versuch einer Entwickelungsgeschichte der Torfmoose* (Sphagnum) *u. einer Monographie der in Europa vorkommenden Arten dieser Gattung.* Stuttgart, 1858. p. 54.
SCHLEIDEN, M. J. (1842). *Grundzüge der wissenschaftlichen Botanik.* 1 Aufl. 1842–43; 2 Aufl. 1845–46; 3 Aufl. 1849–50. Leipzig.
SCHLICHT, A. (1888). Ueber neue Fälle von Symbiose der Pflanzenwurzeln mit Pilzen. *Ber. d. d. bot. Gesell.* **6**, 1888, p. 269.
—— (1889). Beitrag zur Kenntniss der Verbreitung und der Bedeutung der Mykorrhizen. Inaug. Dissert. *Landwirtschaftl. Jahrb.* **18**, 1889, p. 499.
SCHNEIDER, A. (1892). Observations on some American Rhizobia. *Bull. Torrey Bot. Club*, **19**, 1892, p. 203.
—— (1894). Mutualistic symbiosis of Algae and Bacteria with *Cycas revoluta*. *Bot. Gazette*, **19**, 1894, p. 25.
—— (1903). History of leguminous nodules, and literature concerning fixation of free nitrogen by plants. *Minnesota Botanical Studies*, 1903.
SCHRÖTER, C. (1908). *Das Pflanzenleben der Alpen.* Zurich, 1908.
SCHULTZ, M. (1861). Ueber Muskelkörperchen und das was man eine Zelle zu nennen hat. *Arch. Anat. Phys.* 1861.
SCHWARTZ, E. J. (1912). Observations on *Asarum europaeum* and its mycorhiza. *Ann. of Botany*, **26**, 1912, p. 769.
SCHWENDENER, S. (1867). Ueber den Bau des Flechtenthallus. *Verh. Schweiz. Naturforsch. Gess. Aarau*, 1867, p. 88.
SERVATTAZ, C. (1913). Le développement et la nutrition des Mousses en milieux sterilisés. *Ann. d. Sc. nat. Bot.* **17**, 1913, p. 111.
SEWARD, A. C. (1898). *Fossil Plants.* Cambridge Univ. Press, 1898. Vol. 1, pp. 207–222.
SHANTZ, H. L. and PIEMEISEL, R. L. (1917). Fungus fairy rings in E. Colorado and their effect on vegetation. *Journ. Agric. Research*, **11**, 1917, p. 191.
SHIBATA, K. (1902). Cytologische Studien über die endotrophen Mykorrhizen. *Prings. Jahrb. f. wiss. Bot.* **37**, 1902, p. 643.
SHIBATA, K. und TAHARA, M. (1917). Studien über die Wurzelknöllchen. *Bot. Magazine*, Tokyo, **31**, 1917, p. 157.
SIMONET, M. (1925). Les champignons endophytes des Orchidées. *Rev. path. veg. et Ent. agric.* **12**, 1925, p. 204.
SMITH, ERWIN (1917). Mechanism of overgrowth in plants. *Proc. Amer. Philosoph. Soc.* **56**, 1917.
SMOTLACHA, FR. (1911). Monographische Bearbeitung der Boletinen Böhmens (tschechisch). *Sitz.-Ber. Kgl. Böhm. Gesellsch. d. Wiss. math.-naturw. Kl.* 1911. (See also *Centralbl. f. Bakt.* Abt. 2, **42**, 1915.)
SOLMS-LAUBACH, H. (1867–68). Ueber den Bau u. die Entwickelung der Ernährungsorgane parasitischen Phanerogamen. *Prings. Jahrb. f. wiss. Bot.* **6**, 1867–68, p. 509.
—— (1874) Ueber d. Bau d. Samen in d. Familien d. Rafflesiaceae u. Hydnoraceae. *Bot. Zeitung*, 1874, p. 358.

SOLMS-LAUBACH, H. (1884). Der Aufbau des Stockes von *Psilotum triquetrum* u. dessen Entwickelung aus der Brutknospe. *Ann. du Jard. bot. de Buitenzorg*, **4**, 1884, p. 139.

SORAUER, P. (1886). *Handbuch der Pflanzenkrankheiten*. 2 Aufl. Berlin, 1886.

SPESSARD, E. A. (1917). Prothallia of *Lycopodium* in America. *Bot. Gazette*, **63**, 1917, p. 66; **65**, 1918, p. 362; **74**, 1922, p. 392.

SPRATT, E. R. (1912). The formation and physiological significance of the root-nodules in the Podocarpaceae. *Ann. of Botany*, **26**, 1912, p. 801.

—— (1915). The root-nodules of the Cycadaceae. *Ann. of Botany*, **29**, 1915, p. 619.

—— (1919). A comparative account of the root-nodules of the Leguminosae. *Ann. of Botany*, **33**, 1919, p. 189.

SPRING, A. F. (1842). *Monographie de la famille des Lycopodiacées*, 1842.

STAHL, E. (1877). *Beiträge zur Entwickelungsgeschichte der Flechten*. Heft 2. Leipzig, 1877.

—— (1900). Der Sinn der Mycorrhizenbildung. *Jahrb. f. wiss. Bot.* **34**, 1900, p. 539.

STOKEY, A. G. und STARR, A. M. (1924). *Lycopodium* prothallia in western Massachusetts. *Bot. Gazette*, **77**, 1924, p. 80.

STUTZER, A. und KLINGENBERG, W. (1882). Über die Versetzbarkeit stickstoffhaltiger animalischer Düngstoffe. *Journ. Landwirtsch.* **30**, 1882, p. 363.

TERNETZ, C. (1904). Assimilation des atmosphärischen Stickstoffes durch einen torfbewohnenden Pilz. *Vorl. Mitt. Ber. d. d. bot. Gesell.* **22**, 1904, p. 267.

—— (1907). Über die Assimilation des atmosphärischen Stickstoffes durch Pilze. *Jahrb. f. wiss. Bot.* **44**, 1907, p. 353.

THESLEFF, A. (1919). Studier öfver basidsvampfloran i sydöstra Finland med häusyn till dess sammansättung, fysiognomi, fenologi, och ekologi. *Bidr. till kännedom af Finlands Natur och Folk*, **79**, Helsingfors, 1919.

THISTLETON-DYER, W. T. (1897). Note on the discovery of mycorhiza. *Ann. of Botany*, **11**, 1897, p. 175.

THOMAS, A. P. W. (1901). Preliminary account of the prothallium of *Phylloglossum*. *Proc. Roy. Soc. London*, **69**, B, 1901, p. 285.

THOMAS, FR. (1885). Zur Beziehung zwischen Pilzen einerseits und Gallen sowie Gallmückenlarven andererseits. *Irmischia*, 1885, No. **1**, p. 4; abstract in *Bot. Centralbl.* **22**, 1885.

THOMAS, M. B. (1893). The genus *Corallorhiza*. *Bot. Gazette*, **18**, 1893, p. 166.

THORNTON, H. G. and GANGALEE, N. (1926). The life-cycle of the nodule organism, *Bacillus radicicola* (Beij.) in soil, and its relation to the infection of the host plant. *Proc. Roy. Soc. London*, **99**, B, 1926, p. 427.

TREUB, M. (1884–90). Études sur les Lycopodiacées. *Ann. du Jard. bot. de Buitenzorg*, **4**, 1884, p. 107; **5**, 1885–86, p. 87; **7**, 1887–88, p. 141; **8**, 1889–90, p. 1.

TUBEUF, C. F. VON (1888). *Beiträge zur Kenntniss der Baumkrankheiten*. Berlin, 1888.

—— (1895). *Pflanzenkrankheiten*. Stuttgart, 1895. (Eng. ed. Smith, W. G. 1897.)

—— (1896). Die Haarbildung der Coniferen. *Forst. naturwissensch. Zeit.* **5**, 1896.

—— (1903). Beiträge zur Mykorrhizafrage. *Centralbl. f. Bakt. Parasit. u. Infekt.* **10**, 1903, p. 481.

TULASNE, C. (1851). *Fungi hypogaei. Histoire et monographie des champignons hypogés*. Paris, 1851. (Also later ed. Paris, 1862.)

TULASNE, L. R. and C. (1841). Observations sur le genre *Elaphomyces*, et description de quelques espèces nouvelles. *Ann. d. Sc. nat. Bot.* 2me sér. **16**, 1841, p. 5.

UNGER, F. (1840). Beiträge zur Kenntniss der parasitischen Pflanzen. *Ann. des Wiener Museums der Naturgeschichte*, **2**, Wien, 1840, pp. 13–60.

VOGL, A. E. (1898). *Zeitschrift f. Währungsmittel Untersuchungen, Hygiene, u. Waarenkunde*, **12**, 1898.

VUILLEMIN, P. (1889). Antibiose et symbiose (1889). *Association française pour l'avancement des Sciences, Congrès de Paris*, **18**, 1889.

—— (1890). Les mycorhizes. *Revue générale des sciences pures et appliquées*, 1re année, 1890, p. 326.

WAHRLICH, W. (1886). Beitrag zur Kenntniss der Orchideenwurzelpilze. *Bot. Zeitung*, **44**, 1886, p. 480.

WAKSMAN, S. A. (1916). Soil fungi and their activities. *Soil Science*, **2**, 1916.

—— (1924). Influence of micro-organisms upon the carbon-nitrogen ratio in the soil. *Journ. of Agric. Sc.* **14**, 1924, p. 555.

WARD, H. MARSHALL (1887). On the tubercular swellings on the roots of *Vicia faba*. *Phil. Trans. Roy. Soc. London*, **178**, 1887.

—— (1889). On the tubercles on the roots of Leguminous plants, with special reference to the Pea and Bean. *Proc. Roy. Soc. London*, **46**, B, 1889, p. 431.

—— (1899). Symbiosis. *Ann. of Botany*, **13**, 1899, p. 549.

WARMING, E. (1876). Smaa biologiske og morfologiske Bidrag. *Bot. Tidskrift*, **1**, 1876, p. 84. (See also *Jost's Jahresber.* 1876, p. 439.)

WARNSTORF, C. (1886). Zur Frage über die Bedeutung der bei Moosen vorkommenden zweierlei Sporen. *Verhandl. d. bot. Vereins d. Prov. de Brandenburg*, **27**, 1886, p. 181. (See also *Revue Bryologique*, 1887, p. 15.)

WEBER, C. (1884). Ueber den Pilz der Wurzelanschwellungen von *Juncus bufonius*. *Bot. Zeitung*, 1884, p. 369.

WEBER, F. (1920). Notiz zur Kohlensäureassimilation von *Neottia*. *Ber. d. d. bot. Gesell.* **38**, 1920, p. 233.

WEEVERS, T. (1916). Das Vorkommen des Ammoniaks und der Ammonsalze in den Pflanzen. *Recueil des Travaux botaniques Néerlandais*, **13**, 1916, p. 63.

WEISS, F. E. (1904). Mycorrhiza from the Lower Coal Measures. *Ann. of Botany*, **18**, 1904, p. 255.

—— (1916). Seeds and Seedlings of Orchids. *Annual Report and Trans. Manchester Microscop. Soc.* 1916, p. 32.

WEST, C. (1916). *Stigeosporium Marattiacearum*. Gen. et sp. nov. *Ann. of Botany*, **30**, 1916, p. 357.

—— (1917). Mycorrhiza of Marattiaceae. *Ann. of Botany*, **31**, 1917, p. 77.

WEYLAND, H. (1912). Zur Ernährungsphysiologie mykotropher Pflanzen. *Prings. Jahrb. f. wiss. Bot.* **51**, 1912, p. 1.

WILSON, W. (1844). Notes on *Monotropa Hypopitys*. *The Phytologist*, **1**, 1844, p. 148.

WINOGRADSKY, S. (1895). Recherches sur l'assimilation de l'azote libre de l'atmosphère par les microbes. *Arch. des Sc. biolog. de St Pétersbourg*, **3**, 1895, p. 297.

—— (1902). *Centralbl. f. Bakt.* Abt. II, **9**, 1902.

WOLFF, H. (1925). Zur Physiologie des Wurzelpilzes von *Neottia Nidus avis* L. *Verhandlungen d. Schweizer Naturforsch. Gesellsch.* **106**, Jahresversammlung, Aarau, Part 2, 1925, p. 155.

—— (1926). *Zur Physiologie des Wurzelpilzes von* Neottia Nidus avis *und einigen grünen Orchideen*. Inaug. Dissert. Universität Basel, 1926.

WOLFF, J. (1923 *a*). Conditions favorables ou nuisibles à la germination des semences Orchidées et au développement des plantules. *C. R. Acad. des Sc.* **177**, 1923, p. 888.

—— (1923 *b*). Contribution à la connaissance des phénomènes de symbiose chez les Orchidées. *C. R. Acad. des Sc.* **177**, 1923, p. 554.

—— (1925). Observations sur les divers modes de culture des orchidées. *Rev. path. vég. et ent. agric.* **12**, 1925, p. 185.

WORONIN, M. (1866). Ueber die bei der Schwartzerle u. der gewöhnlichen Garten-Lupine auftreten den Wurzelanschwellungen. *Mém. de l'Acad. Imp. des Sc. St Pétersbourg*, **10**, 1866, p. 2.

—— (1867). Observation sur certaines excroissances que présentent les racines de l'Aune et du Lupin des jardins. *Ann. d. Sc. nat. Bot.* 5me sér. **7**, 1867, p. 73.

—— (1885 *a*). Ueber die sogenannte Pilzwurzel (Mykorrhiza) von B. Frank. *Ber. d. d. bot. Gesell.* **3**, 1885, p. 205.

—— (1885 *b*). Bemerkungen zu dem Hufsatze von Herrn H. Möller über *Plasmodiophora Alni*. *Ber. d. d. bot. Gesell.* **3**, 1885, p. 177.

WRIGHT, W. (1925). *Nodule bacteria of Soy bean.* I. *Bacteriology of strains.* II. *Nitrogen fixation experiments.* Soil Sci. **20**, 1925, pp. 95, 131.

WULFF, T. (1902). *Botanische Beobachtungen aus Spitzbergen.* Lund, 1902 (See also *New Phytologist*, **1**, 1902, p. 81.)

YEATES, J. S. (1924). The root-nodules of New Zealand Pines. *N.Z. Journ. Sc. and Tech.* **7**, 1924, p. 121.

ZACH, F. (1910). Studie über Phagocytose in den Wurzelknöllchen der Cycadeen. *Oest. Bot. Zeitschrift*, **60**, 1910, p. 49.

ZELLNER, J. (1923). Die Symbiose der Pflanze als chemisches Problem. *Beih. z. Bot. Centralbl.* **40**, 1923, p. 1.

ZIEGENSPECK, H. (1922). Lassen sich Beziehungen zwischen dem Gehalte an Basen in der Asche und dem Stickstoffgehalte der Pflanzen aufstellen, die einen Rückschluss auf die Ernährungsart und die Exkretion gestatten. *Ber. d. d. bot. Gesell.* **40**, 1922, p. 78.

INDEX

www.ingramcontent.com/pod-product-compliance
Lightning Source LLC
Chambersburg PA
CBHW020528270326
41927CB00006B/492

* 9 7 8 1 5 2 8 7 0 7 6 3 3 *